saur

UDC 001.4::659.2(036)

Infoterm Series 4

Organizations
Commissions
Terminology Banks

World Guide to Terminological Activities

2nd completely revised and enlarged edition

Prepared by
Magdalena Krommer-Benz

K·G·Saur München · New York · London · Paris 1985

Edited by
International Information Centre for Terminology
Centre International d'Information pour la Terminologie
Международный информационный центр по терминологии
Internationales Informationszentrum für Terminologie
Infoterm is affiliated to ON (Austrian Standards Institute)
Leopoldsgasse 4 · A-1021 Wien 2 · Austria

Infoterm, established in 1971, is sponsored by Unesco within the framework of the UNISIST programme. It works in liaison with the Technical Committee 37 "Terminology (principles and co-ordination)" of the International Organization for Standardization (ISO).
Infoterm receives financial support from Bundesministerium für Bauten und Technik, Bundeskammer der gewerblichen Wirtschaft, Österreichisches Normungsinstitut

Produced with the aid of a typesetting programme developed by the Institut für Maschinelle Dokumentation of the Forschungsgesellschaft Joanneum, Graz, Austria

CIP-Kurztitelaufnahme der Deutschen Bibliothek

Krommer-Benz, Magdalena:
World guide to terminological activities :
organizations, comm., terminology banks /
prepared by Magdalena Krommer-Benz. [Ed. by
Internat. Information Centre for Terminology]
. – 2., completely rev. and enl. ed. –
München ; New York ; London ; Paris : Saur,
1985.
(Infoterm series ; 4)
1. Aufl. im Verl. Dokumentation, München
ISBN 3-598-21368-9
NE: Internationales Informationszentrum für
Terminologie <Wien>: Infoterm series; HST

2nd completely revised and enlarged edition.

Printed in the Federal Republic of Germany
by Hain-Druck, Meisenheim/Glan
Bound by Thomas Buchbinderei, Augsburg
ISBN 3-598-21368-9

TABLE OF CONTENTS

PREFACE

As befits an institution that functions as a focal point within the framework of the International Network for Terminology (TermNet), the International Information Centre for Terminology (Infoterm), in response to the mandate given to it by the United Nations Educational, Scientific and Cultural Organization (Unesco), co-ordinates activities for collecting, analysing, recording, processing and disseminating data and information on terminology.

Established in 1971, Infoterm collects data on institutions, organizations and individuals working on terminologies. This edition of the **World Guide to Terminological Activities,** which updates the first edition published in 1977, presents the computerized data in a new format. It is hoped that the **Guide** will be a useful tool for those seeking information on who is doing what and where. Mention should be made, however, of the fact that the data presented can be only as complete and up-to-date as the data returned to Infoterm. It is envisaged that TermNet partners and others engaged in terminology work should take over the data collection for specific geographical or subject areas to ease the work of Infoterm in updating this reference tool for the benefit of all concerned.

ACKNOWLEDGEMENTS

Special thanks are due to all who made the collection of the data possible, particularly

— the General Information Programme of Unesco for financial support of data collection and processing;
— the many respondents who supplied data;
— the Institut für maschinelle Dokumentation, Graz, Austria (expert in charge Dipl.-Ing. F. Klemm) for processing the multilingual data and preparing it for phototype setting;
— the Infoterm staff for valuable assistance.

Vienna, November 1984

Prof. Helmut Felber
Director of Infoterm

INTRODUCTION

This completely revised and expanded version of the World Guide provides comprehensive coverage of terminological activities carried out by organizations, commissions/committees and data banks worldwide. The need for such a reference tool became particularly evident with the establishment of TermNet in 1979.

TermNet comprises three separate programmes

- Programme 1: Developing the scientific basis for terminology (General theory and principles of terminology)
- Programme 2: Establishing closer co-operation in preparing terminologies
- Programme 3: Establishing closer co-operation in collecting, recording, processing and disseminating terminological data and information.

The work is part of Programme 3 as direct communication and exchange of terminological data are essential for the further development and implementation of TermNet.

Data collection

The data were collected between 1981 and 1982 with the aid of a set of four questionnaires specifically designed for computer input (see Appendix). Initially, the questionnaires were sent to some 500 institutions known to be involved in terminology. Nearly 80 per cent of the questionnaires were completed representing a high degree of interest on the part of respondents. The institutions thus addressed were asked to provide information on their own terminology work; in addition, specific questionnaires were used to collect data on terminology work carried out by commissions/committees and by terminological data banks. A distinction between organizations and commissions/committees has been made as the latter are primarily project-oriented.

The information obtained was then processed and stored on floppy disks; due to the lack of computer equipment on Infoterm premises, the actual data processing was carried out by the *Institut für maschinelle Dokumentation*, Graz, Austria. In co-operation with the *Institut*, Infoterm also undertook extensive preparatory work for computer-directed typesetting. Compared to the first edition, this new typeface will facilitate use of the *Guide* since important elements within an entry (such as name, headings, etc.) are set in boldface and can be readily retrieved. The work involved in storing the data on computer will assist updating and publishing of future editions of the *Guide*.

Arrangement of entries

The arrangement of entries is based on the Universal Decimal Classification (UDC) notations of countries. A list of countries in alphabetical order, with corresponding UDC numbers *(Geographic Index)* is provided on p. 1 for those unacquainted with the system. Within each country individual organizations, commissions/committees and data banks are listed by name in a single alphabetical sequence. Commissions/committees or data banks affiliated to a specific organization and their projects are listed in alphabetical order immediately after the parent organization. All main entries are numbered consecutively.

Contents of entries

The information presented in the *Guide* is that provided by the respondents who are also responsible for the accuracy and reliability of data. For this reason the length and substance of entries vary according to the amount and type of information received. Between 1982 and 1983, all respondents were contacted and asked to verify and/or supplement the draft entries. An exception was made for a number of standards bodies all of which standardize the terminologies for a variety of fields in the respective countries and languages.

Special symbols are used to classify organizations, commissions/committees and data banks involved in terminology work as follows:

● Those primarily involved in terminology work;
◐ Those marginally involved in terminology work;
○ Those known to be carrying out terminology work but for which no data are available.

In principle, the contents of the entries are self-explanatory as headings are provided for all data supplied. For purposes of simplification, the information headed by "Hardware/Capacity" contains data on the following items, if provided by the respondents:

Make/model/memory/storage/peripherals/terminals

For projects, special information under the heading "Responsible" is only given if the implementing agency is different from "Principal investigator". The language codes used (see p. 2) are those recommended by the International Organization for Standardization (ISO).

Indexes

A cumulative alphabetical list of all organizations, commissions/committees and data banks is given on p. 116 — p. 134 listing original name, translation(s), abbreviations(s) and acronym(s) provided by the respondents. Names and abbreviations/acronyms in the original language are given in boldface. Special indexes of commissions/committees and data banks are provided for those seeking more specific information. A comprehensive list of abbreviations and acronyms appearing in the *Guide* is given on p. 151. A subject index is included on p. 148. Unless indicated otherwise, indexes refer to the numbers of the entries.

Information available via data base only

For the purposes of the present *Guide* no detailed information on holdings, internal structure, co-operation with other institutions, unpublished manuscripts, etc. is given. Infoterm is completing arrangements for such information to be available for on-line queries via data base hosts. Parties interested in on-line access will be informed when this system is operational.

Although the utmost care has been taken in the preparation of the *Guide*, Infoterm would welcome comments, suggestions and possible corrections from users of the *Guide*.

Magdalena Krommer-Benz, M. A.
Infoterm

GEOGRAPHIC INDEX

LANGUAGE CODE [1]

Af	Afrikaans
Ar	Arabic
Be	Byelorussian
Bg	Bulgarian
Ch	Chinese
Cr	Croatian
Cs	Czech
Da	Danish
De	German
En	English
Eo	Esperanto
Es	Spanish
Et	Estonian
Ew	Ewe
Fa	Farsi
Fi	Finnish
Fr	French
Gr	Greek
He	Hebrew
Hi	Hindi
Hu	Hungarian
Ia	Interlingua
Ie	Interlingue (= Occidental)
In	Indonesian
Is	Icelandic
It	Italian
Ja	Japanese
Ko	Korean
La	Latin
Le	Lettish, Latvian
Li	Lithuanian
Mc	Macedonian
Nl	Dutch
No	Norvegian
Pi	Pilipino
Pj	Punjabi
Pl	Polish
Pt	Portuguese
Ro	Romanian
Ru	Russian
Sa	Sanskrit
Sh	Serbo-Croat
Sk	Slovak
Sn	Slovenian
Sr	Serbian
Sv	Swedish
Sw	Swahili
Th	Thai
Tk	Turkmen
Tr	Turkish

[1]) Based on ISO/DP 639 (1979) and ISO/DP 639 (1982). The forthcoming standard (ISO 639) will prescribe minuscules only.

0/9	Generalities. All fields
0/6	Generalities. Philosophy. Social sciences. Law, mathematics. Natural sciences. Applied sciences. Medicine. Technology
001.4	Specialist terminology. Scientific nomenclature
001.811	Rules for citations: bibliographic references
002	Documentation. Books. Writings. Authorship
007	Information. Communication and control theory (cybernetics)
008	Civilization. Culture. Progress
016	Special subject bibliographies
02	Librarianship
025.4	Classification. Indexing languages
026	Special libraries
07	Newspapers. Journalism
1	Philosophy
159.9	Psychology
2	Religion
3/7	Social sciences. Mathematics and natural sciences. Applied sciences. Medicine. Technology. The Arts
3/6	Social sciences. Philology. Mathematics and natural sciences. Applied sciences. Medicine. Technology
3	Social sciences. Law. Government
31	Statistics. Demography. Population studies
311	Statistics as a science. Statistical theory
32	Politics
33	Political economy. Economics
330	General concepts of economics
331.01	Theory of economic labour
336	Finance. Public finance. Customs, banking and monetary system
336.2	Taxes. Dues. Fees. Payments
336.71	Banks and banking
338	Production of wealth. Production. Trend of the economy. Economic situation
339	Trade. Commerce
34	Law. Legislation
347.8	Air law. Broadcasting law
35	Administration
351	Spheres of activity of public administration
351.712	Public works. Contracts with contractors
351.816	Post. Postal administration law. Postal law
351.84	Welfare work. Public assistance. Social insurance
356	Infantry and foot troops generally
36	Social welfare and relief
362.1	Medical attention and aid. Needy (Family welfare)
368.4	Social insurance
37	Education. Teaching. Training. Leisure
37.014	Educational policy. Educational work and public life
372	Content, curriculum in pre-school and school education. Subjects (at any level and any type of school)
5/7	Mathematics and natural sciences. Applied sciences. Medicine. Technology. The Arts. Architecture. Entertainment
5/6	Mathematics and natural sciences. Applied sciences. Medicine. Technology
5	Mathematics and natural sciences
51	Mathematics
519.2	Probability and mathematical statistics
52	Astronomy. Astrophysics. Space research. Geodesy
528.9	Cartography. Mapping
53	Physics
532	Mechanics of fluids in general. Mechanics of liquids
534	Acoustics. Vibrations
539.1	Nuclear, atomic and molecular physics
54	Chemistry. Crystallography. Mineralogy
541.1/.4	Physical chemistry. Atomic theory. Chemical combination. Classes of chemical compounds
541.64	Macromolecular chemistry. Polymerization products
542	Practical laboratory chemistry, preparative and experimental
543	Analytical chemistry
564	Inorganic chemistry
547	Organic chemistry
548	Crystal science. Crystallography
55	Earth sciences. Geology. Geophysics. Seismology. Meteorology. Hydrology
551	General geology. Meteorology. Climatology. Historical geology. Stratigraphy. Palaeogeography
551.326	Floating ice
551.5	Meteorology
551.7	Historical geology. Stratigraphy
556	Hydrosphere. Water in general. General hydrology
57	Biological sciences
574	General ecology. Hydrobiology. Biogeography
575	General genetics. General cytogenetics. Immunogenetics. Evolution. Speciation. Phylogeny
576	Cellular and subcellular biology. Cytology
576.8	Parasitology
577	Material bases of life. Biochemistry. Molecular biology. Biophysics (General studies of the chemical and physical-chemical foundations of biological phenomena)
577.1	Chemical bases of life. Biochemistry and bioorganic chemistry generally
578	Virology
58	Botany
581.46	Flowers
59	Zoology
6	Applied sciences. Medicine. Technology
60	General questions of the applied sciences
61	Medicine
612	Physiology
612.39	Nutrition
614	Public health. Accident prevention
614.7	Hygiene of the air, water and soil
614.8	Accidents. Risks of hazards. Prevention, personal protection and safety
615	Pharmaceutics. Therapeutics. Toxicology
615.4	Pharmaceutical preparations. Medical materials. Equipment
616-089	Operative treatment. Operative technique. Surgical, operative procedure. Surgery
616.314	Teeth. Odontology
62	Engineering. Technology in general
620.1	Materials testing. Defects of materials. Protection of materials
620.1.05	Testing machines and apparatus
620.17	Measurement of mechanical properties of materials. Strength tests
620.179.1	Non-destructive testing
620.179.11	Surface tests
620.193	Physical and chemical influence. Corrosion
620.2	Commercial materials. Goods. Wares
620.9	Economics of energy in general
621-752	Means of damping shock, oscillation, vibration
621.039	Applied nuclear science. Atomic energy and atomic industry. Nuclear engineering in general
621.22	Hydraulic energy. Water power. Hydraulic machinery
621.3	Electrical engineering
621.38	Electronic devices. Electron tubes. Photocells. Particle accelerators. X-Ray tubes
621.39	Telecommunication. Telegraphy. Telephony. Radiocommunication. Facsimile. Television. Telecontrol
621.43	Internal combustion engines
621.5	Pneumatic energy, machinery and tools. Refrigeration and liquefaction techniques and plant

621.6.02	Gases. Vapours. Air
621.7	Plastic or chipless working, forming and treatment of materials in general: processes, tools, machines, equipment
621.762	Powder metallurgy
621.78	Heating for working. Heat treatment. Hardening
621.791	Welding and allied techniques
621.793/.795	Metal coating and plating. Production of conducting, semiconducting, resistive, dielectric, magnetic coats and films. Chemical finishing and surface treatments. Final surface treatment: decoration, colouring, marking, etc.
621.798	Packaging and packaging equipment: containers, materials, machines, etc.
621.822	Bearings. Bushes
621.83	Mechanical transmissions. Gears, gearing. Cams. Sliding tracks. Ratchets. Clutches
621.867	Continuous transfer equipment. Conveyors. Elevators. Chutes
621.868.2	Free-ranging floor and yard vehicles
621.873	Cranes in general
621.879.4	Machines acting continuously
621.9	Working, machining with chip formation. Cutting, grinding, sheet-working, etc.
622	Mining
622.279	Extraction of (natural) gas
623	Military engineering
624.04	Structural design. Graphical and analytical statistics for investigation and calculation of structures
624.136	Embankments. Coffer dams. Dykes
625.7/.8	Highway engineering
626	General hydraulic engineering. Canal engineering. Agricultural hydraulics
627.133	Gauging of watercourses (quantity and velocity measurements)
627.5/.7	Other structures and works for maintenance and improvement of waterways and coasts. Navigation, channel maintenance, marine salvage and facilities
628	Public health engineering. Water. Sanitation. Illuminating engineering
629	Transport vehicle engineering
629.11	Land vehicles (except rail vehicles). Road vehicles
629.11.012	Suspension. Running gear
629.118	Vehicles with one or two wheels
629.12	Water transport. Water craft. Shipbuilding. Boatbuilding
629.7	Air and space transport engineering. Aeronautics, aircraft. Rockets and missiles. Astronautics, spacecraft
629.73	Aeronautical engineering. Aircraft
63	Agriculture and related sciences and techniques. Forestry. Farming. Wildlife exploitation
630	Forestry
631	Agriculture in general
631.3	Agricultural machines, implements and equipment
631.8	Fertilizers and manures. Growth substances
632.95	Plant protection preparations. Pesticides
633/637	Specific crops. Horticulture. Animal breeding. Produce of domestic animals and game.
633.1	Cereals. Grain crops
634	Horticulture generally
639.2	Fishing. Fishing produce
64	Home economics
641	Foodstuffs. Cooking
645.4	Furniture. Furniture design and manufacture
65	Management and organization in industry, business, trade, commerce, communication and transport
651.5	Office filing. Records, dossiers, etc.
654	Telecommunication and telecontrol (organization, services)
654.1	Telegraphy. Telephony. Radio. Television
655.1/.3	Printing
656	Transport and postal services. Traffic organizations and control
656.07	Administration, management of traffic operations
656.1	Road transport
656.7	Air transport
656.8	Postal services and administration
657.1	Bookkeeping systems and methods
658.56	Quality control. Tool and material control
658.8	Marketing. Selling. Distribution
659	Publicity. Advertising. Public relations
659.2	Information. Advisory and consultancy services
659.3	Mass communication. Education of the public
66	Chemical technology. Chemical and related industries
66.07	Handling, processing and purification of gases
662.6	Fuel economy. Fuels. Heating
662.9	Furnace and combustion engineering
662.998	Industrial heat insulation and insulators. Heat loss
663/665	Beverages. Stimulants. Narcotics. Manufacture and preservation of solid foods. Oils. Fats. Waxes. Adhesives. Gums. Resins
663	Beverages. Stimulants. Narcotics
664.2	Starch. Starch-containing products
664.6/.7	Technology and processing of cereal grains (milling, baking, etc.)
665.5	Essential oils. Perfumery. Cosmetics
665.7	Products of the mineral oil industry. Mineral oils and similar products
666	Glass and allied industries. Ceramics, clay and allied industries. Gypsum, lime, cement and allied industries
666.17	Hollow glass in general without reference to method of manufacture. Blown hollow glass in particular
666.76	Refractory bodies and products
667.6	Coatings and coating techniques. Paints. Varnishes. Lacquers
668.5	Essential oils for various purposes. Perfumery. Cosmetics
669	Metallurgy
669.018.8	Corrosion-resistent alloys
669.14	Alloys of iron with carbon (other than cast iron). Steel in general
669.2/.8	Non-ferrous metallurgy. Non-ferrous metals
669.24	Nickel. Nickel alloys
669.3	Copper. Copper alloys
669.71/.72	Aluminium. Magnesium. Beryllium
67	Various industries, trades and crafts
674.05	Woodworking tools. Woodworking machinery
674.06	Products of the timber and woodworking industry
675	Leather industry
676	Pulp and paper industry. Production of pulp, paper and board
677	Textile industry
677.05	Textile machinery and equipment
677.721	Steel wire ropes and cables
678.4	Natural rubber and other natural macromolecular materials containing carbon and hydrogen only
678.5/.8	Plastics, especially semi-synthetic plastics based on cellulose, proteins and other natural macromolecular materials containing carbon, hydrogen and oxygen or other elements. Synthetic polycondensation products. Polycondensates. Synthetic resins. Synthetic fibres. Synthetic polymerization products. Polymerizates. Synthetic rubbers. Various macromolecular materials.
68	Industries, crafts and trades for finished or assembled articles

681.11	Horology. Watch and clock making
681.3	Data processing equipment
681.6	Graphic reproduction machines and equipment
685.6	Travel, sports, games and other equipment
687	Clothing industry. Articles of toilet
687.053	Sewing machines
69	Building trade. Building materials. Building practice and procedure
691.32	Concrete. Concrete building elements
692.8	Doors. Gates. Windows. Service openings
697	Heating, ventilating and air conditioning of buildings
699.81	Fire protection. Fireproofing of buildings. Firefighting installations and equipment
711	Principles and practice of physical planning. Regional, town and country planning
711.3	Country planning. Rural development
72	Architecture
744.4	Drawing methods, rules, conventions
778.14	Microcopy(ing). Microphotographic technique and equipment, including reading apparatus
79	Recreation. Entertainment. Amusements. Pastimes
792	Theatre. Stage craft. Dramatic performances
80	Linguistics. Philology. Languages
801	General linguistics and philology
801.3	Lexicology. Vocabulary. Word and name research. Lexicography. Dictionaries
801.316	Various categories of words
801.316.1	Neologisms
801.316.4	Technical and other special subject words
804.0	French
82.03	Interpretation. Translation
9	Geography. Biography. History
91	Geography. Exploration of the earth and of individual countries. Travel. Regional geography

● *1* *(Organization)*

Association internationale de traitement thermique des matériaux / International Federation for the Heat Treatment of Materials

[Internationaler Verband für die Wärmebehandlung der Werkstoffe (De); Meždunarodnoe obščestvo po termičeskoj obrabotke materialov (Ru)]

AITT/IFHT; IVW (De); MOTOM (Ru)

P. O. Box 175
Zürich, CH-8047
Switzerland

Staff: 14 (committee members working on voluntary basis)
Head: M. Guy Murry, Paris, president until Dec. 31, 1980; Prof. Dr. Ing. Tibor Konkoly, Budapest, president from Jan. 1, 1981, onwards
Expert(s): Prof. Dr. hab. Eugeniusz Tyrkiel, Warsaw, chairman of Committee "Terminology"
Staff categories:
14 terminologist(s) (theory)
Founded: 1971
Documentation:
bibliographic data
factographic data
Subject(s): heat treatment (of materials)
UDC 621. 78
Language(s): Fr; En; De; Ru; Bg; Es; Fi; Hu; It; Ja; Nl; Pl; Sh; Sv
Vocabularies:
Subject(s): heat treatment (of materials)
UDC 621. 78
Language(s): Fr; En; De; Ru; Bg; Es; Fi; Hu; It; Ja; Nl; Pl; Sh; Sv
Publications: The individual chapters of the Glossary under elaboration appear successively in some national journals, e. g. in Poland: *Metaloznawstwo i obróbka cieplna*, 1979, No. 40, No. 41, No. 42; 1980, No. 44, No. 45; in Germany: *Härterei - Technische Mitteilungen*, 1978, No. 6; 1980, No. 2; in Spain: *Técnica metalúrgica*, 1979, vol. 34, No. 231, No. 232. The publication in this form (for public discussion) is being continued.
Terminological data:
definitions (in Fr/En/De/Ru); equivalents

(Project(s))
Responsible: Technical Committee "Terminology" of the IFHT
Title: Glossaire multilingue de traitement thermique des matériaux [Multilingual Glossary of Heat Treatment of Materials]
Project type: institutional; in progress
Principal investigator: Prof. Dr. hab. Eugeniusz Tyrkiel, Warsaw (Poland)
Collaborators: 14
Scope and/or subject matter: Heat treatment (of materials)
Method of work: Working meetings of the Committee "Terminology" of the IFHT (twice a year); discussion on individual terminological entries
Source material: 1. National standards; 2. Technical literature
Publications: The Glossary will appear in book-form; first volume probably in 1983

● *2* *(Commission/Committee)*

Commission Technique "Terminologie"/Technical Committee "Terminology"

[Fachausschuß "Terminologie"(De); Kommissija "Terminologija" (Ru)]

Parent organization: Association internationale de traitement thermique des matériaux/International Federation for the Heat Treatment of Materials
P. O. Box 175
Zürich, 8047
Switzerland
Contact: Prof. Dr. hab. Eugeniusz Tyrkiel, chairman
Duchnicka 3
PL-00967 Warszawa
Poland
Committee members: 14
Chairperson: Prof. Dr. hab. Eugeniusz Tyrkiel, Warsaw (Poland)
Established: 1974
Vocabularies:
Subject(s): heat treatment (of materials)
UDC 621. 78
Language(s): Fr; En; De; Ru; Bg; Es; Fi; Hu; It; Ja; Nl; Pl; Sh; Sv
Publications: Individual chapters of the Glossary under elaboration appear successively in some national jornals, e. g. in Poland: *Metaloznawstwo i obróbka cieplna*, 1979, No. 40, No. 41, No. 42; 1980, No. 44, No. 45; in Germany: *Härterei - Technische Mitteilungen*, 1978, No. 6; 1980, No. 2; in Spain: *Técnica metalúrgica*, 1979, vol. 34, No. 231, No. 232

● *3* *(Organization)*

Association internationale permanente des congrès de la route/Permanent International Association of Road Congresses

A. I. P. C. R. /P. I. A. R. C.

2, Boulevard de La Tour
Maubourg
Paris, F-75007
France
Phone: 550. 32. 24
Telex: 250857 OFFINAVI Paris, attn. M. Huet
Telegr.: CONGREROUT PARIS

Staff: 6
Head: M. Maurice Milne (United Kingdom), President
Expert(s): M. Roger Coquand (France), Past President
Staff categories:
1 terminologist(s) (theory); 1 translator(s); members of 11 Technical Committees co-operating on questions of terminology
Founded: 1909
Documentation: factographic data
Subject(s): road technology
UDC 625. 7
Language(s): Fr; En
Publications: AIPCR. *Technical Dictionary of Road Terms, English-French.* 4th ed. Paris: AIPCR, 1978, 125 p., 150x235. New edition in preparation and will also be published in a number of booklets.
Services: reference
No service charges
Operation:
manual
Data available:
free of charge

● *4* *(Organization)*

Bureau international pour la standardisation de la rayonne et des fibres synthétiques

[The International Bureau for the Standardisation of Man-Made Fibres]

BISFA

Lautengartenstraße 12
Basel, CH-4010
Switzerland
Phone: 61/23 62 50
Telex: 64 331
Telegr.: BISFA
Head: Dr. Hans Lucas Sarasin, General Secretary
Founded: 1928
Documentation: bibliographic data
Subject(s): man-made fibres
UDC 677
Language(s): Fr; De; En; Es; It; Nl
Vocabularies:
Subject(s): man-made fibres
UDC 677
Language(s): Fr; De; En; Es; It; Nl
Publications: Terminology relating to Man-Made Fibres

● *5* *(Organization)*

Collège international pour l'étude scientifique des techniques de production mécanique

[International Institution for Production Engineering Research; Internationale Forschungsgemeinschaft für mechanische Produktionstechnik]

CIRP
19, rue Blanche Phone: 526 21 80
Paris, F-75009
France
Head: M. F. Le Maitre, Secretary General
Founded: 1951
Vocabularies:
Subject(s): machining; working of materials
UDC 621. 7; UDC 621. 9
Language(s): En; Fr; De; Es; It; Pt; Sv; No; Da
Publications: CIRP. *Dictionary of Production Engineering.* Essen: Verlag W. Girardet. Volume 1 *Forging and Drop Forging* (De - En - Fr) Volume 1. *R Forging and Drop Forging* (De - Es - It - Pt) Volume 1. *N Forging and Drop Forging* (Sv - No - Da)
Volume 2 *Grinding - Surface roughness* (De - En - Fr)
Volume 3 *Sheet metal Forming* (De - En - Fr) Volume 4 *Fundamental Terms of Cutting* (De - En - Fr)
Volume 5 *Cold extrusion and Upsetting* (De - En - Fr)
Volume 6 *Planing, Slotting, Broaching, Turning* (De - En - Fr)
Volume 7 *Drilling, Countersinking/Counterboring,. Reaming, Thread Production*
Volume 8 *Milling, Sawing, Gear Manufacturing* (De - En - Fr). Other volumes are in preparation

● *6* *(Commission/Committee)*
Comité du dictionnaire, du glossaire et du registre mondial des barrages
[Committee on the Dictionary, the Glossary and the World Register of Dams; Comité del diccionario, del glosario y del registro mundial de presas (Es); Comitato del dizionario, del glossario e del registro mondiale delle Dighe (It)]
Parent organization: International Commission on Large Dams (ICOLD) / Commission Internationale des Grands Barrages (CIGB)
151, Boulevard Haussmann Phone: 764–68–24; 764–67–33
Paris, F-75008 Telex: 641 320 F
France (Secretary General, Central Office)
Contact: J. G. du Plessis
Department of Water Affairs
Private Bag X 313
Pretoria 0001
Republic of South Africa
Committee members: 14
Chairperson: J. G. du Plessis
Established: 1976
Vocabularies:
Subject(s): dams
UDC 624. 136
Language(s): En; Fr; De; Es; It; Pt
Vocabulary standards:
Subject(s): dams
UDC 624. 136
Language(s): En; Fr; De; Es; It; Pt
Publications: *Technical Dictionary on Dams,* Appendix A *Glossary of Words and Phrases related to Dams,* published by the International Commission on Large Dams, edition 1978, obtainable from Secretary General

● *7* *(Commission/Committee)*
Commission de terminologie / Commission on Terminology
COMTERM
Parent organization: Association internationale de linguistique appliquée / International Association of Applied Linguistics
Vrije Universiteit Bruxelles, Pleinlaan 2
Bruxelles, 1050
Belgium
Contact: Prof. Dr. L. Drozd, Katedra anglistiky, germanistiky a nordistiky, Filozofická Fakulta University Karlovy, Náměsti Krasnoarméjců 2, CS- 116 38 Praha, CSSR
Committee members: appr. 50 commission members
Chairperson: Prof. Dr. L. Drozd, Prague
Vice-chairperson: Mr. H. Picht, Copenhagen
Established: 1978 (formerly Commission for Terminology, Lexicology and Lexicography)
Theory: Language independent
Publications: COMTERM. Bulletin de la Commission de terminologie de l'Association internationale de linguistique appliquée (irregular)

● *8* *(Organization)*
Commission électrotechnique internationale, Comité d'études no. 1/International Electrotechnical Commission, Technical Committee No. 1
CEI/CE 1; IEC/TC1
Avenue du Général de Gaulle Phone: 33 (1) 765 43 21
Clamart, F-92140 Telex: EDFERIM 270400 F
France
Staff: 7
Expert(s): Mrs. Michelle Hue
Staff categories:
1 terminologist(s) (special field); 2 terminologist(s) (theory); 1 documentalist(s)/librarian(s); 3 other professional(s) (clerical)
Founded: 1910
Documentation: factographic data
Subject(s): electronics; electrical engineering; telecommunications
UDC 621. 3; UDC 654
Language(s): Fr; En; Ru; De; Es; It; Nl; Pl; Sv
Vocabularies:
Subject(s): electronics; electrical engineering; telecommunications
UDC 621. 3; UDC 654
Language(s): Fr; En; Ru; De; Es; It; Nl; Pl; Sv
Publications: IEC/CEI. *International Electrotechnical Vocabulary IEV,* with terms in Fr, En, Ru, De, Es, Nl, Pl, Sv and definitions in Fr, En, Ru. First published in 1938. 2nd edition 1949–1970 covering appr. 8 500 concepts. No 3rd complete edition is envisaged but - based upon need - different chapters are being revised, updated and published.
IEC/CEI. *International Electrotechnical Vocabulary. General index/Vocabulaire électrotechnique international. Index général.* Genève: IEC/CEI, 744 p., 180x260. The index contains terms in En and Fr pertaining to electronics, electrical engineering and telecommunication. Each term appears with an equivalent in the other language and can be found from any of its significant constituent words. The vocabulary covers the terminology up to 1975 and an updated edition was published in 1979. 1011 pages inclusive of symbols.
Services: reference; reproduction; referral
Service charges

(Project(s))
Responsible: Secretariat of IEC, TC 1
Title: Analyse automatique de définitions [Automatic analysis of definitions]
Project type: commissioned project; in progress
Principal investigator: Secretariat of IEC, TC1

Responsible: Secretariat of IEC, TC 1
Title: Elaboration automatique d'un dictionnaire multilingue de l'électricité, de l'électrotechnique et des télécommunications [Automatic setting up of a multilingual dictionary of electricity, electrical engineering and telecommunications]
Project type: commissioned project; in progress
Principal investigator: Secretariat of IEC, TC1
Scope and/or subject matter: Electricity; electrical engineering; telecommunications
Method of work: Computerised method
Source material: IEV publications

● *9* *(Commission/Committee)*
JCG/WG^A "Oscillations-Signals"
Parent organization: International Electrotechnical Commission

Contact: Comité électrotechnique français Paris Cedex, 16 12, Place des Etats-Unis, F-75783, France
Vocabularies:
Subject(s): electrical engineering
UDC 621. 3
Language(s): Fr; En; Ru; De; Es; It; Nl; Pl; Sv

● ***10*** *(Commission/Committee)*
JCG/WG^A "Types of telecommunications and basic terms"
Parent organization: International Electrotechnical Commission
Paris Cedex 16 12, Place des Etats-Unis, F-75783, France
Contact: Comité électrotechnique français
Vocabularies:
Subject(s): electrical engineering
UDC 621. 3
Language(s): Fr; En; Ru; De; Es; It; Nl; Pl; Sv

● ***11*** *(Commission/Committee)*
JCG/WG^B "Electrical networks and network elements"
Parent organization: International Electrotechnical Commission
Contact: Comité électrotechnique français Paris Cedex 16 12, Place des Etats-Unis, F-75783, France
Vocabularies:
Subject(s): electrical engineering
UDC 621. 3
Language(s): Fr; En; Ru; De; Es; It; Nl; Pl; Sv

● ***12*** *(Commission/Committee)*
JCG/WG^C "Transmission and communication techniques"
Parent organization: International Electrotechnical Commission
Contact: UK British Electrotechnical Committee
British Standards Institution London
2, Park Street, W1A 2BS, U. K.
Vocabularies:
Subject(s): electrical engineering
UDC 621. 3
Language(s): Fr; En; Ru; De; Es; It; Nl; Pl; Sv

● ***13*** *(Commission/Committee)*
JCG/WG^D "Waves and propagation"
Parent organization: International Electrotechnical Commission
Contact: Comité électrotechnique français Paris Cedex 16, 12, Place des Etats-Unis, F-75783, France
Vocabularies:
Subject(s): electrical engineering
UDC 621. 3
Language(s): Fr; En; Ru; De; Es; It; Nl; Pl; Sv

● ***14*** *(Commission/Committee)*
JCG/WG^E "Electrical and electronic subassemblies"
Parent organization: International Electrotechnical Commission
Contact: Comité électrotechnique français Paris Cedex 16, 12, Place des Etats-Unis, F-75783, France
Vocabularies:
Subject(s): electrical engineering
UDC 621. 3
Language(s): Fr; En; Ru; De; Es; It; Nl; Pl; Sv

● ***15*** *(Commission/Committee)*
JCG/WG^F "Antennas"
Parent organization: International Electrotechnical Commission
Contact: U. S. National Committee of the IEC, American National Standards Institute Inc., New York, N. Y. 10018, USA 1430 Broadway
Vocabularies:
Subject(s): electrical engineering
UDC 621. 3
Language(s): Fr; En; Ru; De; Es; It; Nl; Pl; Sv

● ***16*** *(Commission/Committee)*
JCG/WG^G "Radiocommunications, radio transmitters and receivers"
Parent organization: International Electrotechnical Commission
Contact: Comité électrotechnique français, Paris Cedex 16, 12, Place des Etats-Unis, F-75783, France
Vocabularies:
Subject(s): electrical engineering
UDC 621. 3

● ***17*** *(Commission/Committee)*
JCG/WG^H "Switching"
Parent organization: International Electrotechnical Commission
Contact: Comité électrotechnique suisse, Association suisse des électriciens, Zürich, Seefeldstraße 301, CH-8034, Switzerland
Vocabularies:
Subject(s): electrical engineering
UDC 621. 3
Language(s): Fr; En; Ru; De; Es; It; Nl; Pl; Sv

● ***18*** *(Commission/Committee)*
JCG/WG^J "Teletraffic-trunking and operating"
Parent organization: International Electrotechnical Commission
Contact: Svenska Elektriska Kommissionen, Stockholm 5, Box 5177, S-102 44, Sweden
Vocabularies:
Subject(s): electrical engineering
UDC 621. 3
Language(s): Fr; En; Ru; De; Es; It; Nl; Pl; Sv

● ***19*** *(Commission/Committee)*
JCG/WG^K "Radio networks and operating"
Parent organization: International Electrotechnical Commission
Contact: Comité électrotechnique français, Paris Cedex 16, 12, Place des Etats-Unis, F-75783, France
Vocabularies:
Subject(s): electrical engineering
UDC 621. 3
Language(s): Fr; En; Ru; De; Es; It; Nl; Pl; Sv

● ***20*** *(Commission/Committee)*
JCG/WG^L "Tests and measuring maintenance and protection of telecommunications networks"
Parent organization: International Electrotechnical Commission
Contact: Comité électrotechnique français, Paris Cedex 16, 12, Place des Etats-Unis, F-75783, France
Vocabularies:
Subject(s): electrical engineering
UDC 621. 3
Language(s): Fr; En; Ru; De; Es; It; Nl; Pl; Sv

● ***21*** *(Commission/Committee)*
JCG/WG^M "Telegraphy and data transmission"
Parent organization: International Electrotechnical Commission
Contact: Comité électrotechnique français, Paris Cedex 16, 12, Place des Etats-Unis, F-75783, France
Vocabularies:
Subject(s): electrical engineering
UDC 621. 3
Language(s): Fr; En; Ru; De; Es; It; Nl; Pl; Sv

● ***22*** *(Commission/Committee)*
JCG/WG^N "Telephony"
Parent organization: International Electrotechnical Commission
Contact: Svenska Elektriska Kommissionen, Stockholm 5, Box 5177, S-102 44, Sweden
Vocabularies:
Subject(s): electrical engineering
UDC 621. 3
Language(s): Fr; En; Ru; De; Es; It; Nl; Pl; Sv

● *23* *(Commission/Committee)*
JCG/WG^O "Optical fibres"
Parent organization: International Electrotechnical Commission
Contact: Vacant
Vocabularies:
Subject(s): electrical engineering
UDC 621. 3
Language(s): Fr; En; Ru; De; Es; It; Nl; Sv

● *24* *(Commission/Committee)*
JCG/WG^O "Transmission lines and waveguides"
Parent organization: International Electrotechnical Commission
Contact: Comité électrotechnique français, Paris Cedex 16, 12, Place des Etats-Unis, F-75783, France
Vocabularies:
Subject(s): electrical engineering
UDC 621. 3
Language(s): Fr; En; Ru; De; Es; It; Nl; Pl; Sv

● *25* *(Commission/Committee)*
JCG/WG^R "Broadcasting services, sound broadcasting, television"
Parent organization: International Electrotechnical Commission
Contact: Comité électrotechnique français, Paris Cedex 16, 12, Place des Etats-Unis, F-75783, France
Vocabularies:
Subject(s): electrical engineering
UDC 621. 3
Language(s): Fr; En; Ru; De; Es; It; Nl; Pl; Sv

● *26* *(Commission/Committee)*
JCG/WG^T "Space radiocommunications"
Parent organization: International Electrotechnical Commission
Contact: Comité électrotechnique français, Paris Cedex 16, 12, Place des Etats-Unis, F-75783, France
Vocabularies:
Subject(s): electrical engineering
UDC 621. 3
Language(s): Fr; En; Ru; De; Es; It; Nl; Pl; Sv

● *27* *(Commission/Committee)*
JCG/WG^Y "Reliability and availability"
Parent organization: International Electrotechnical Commission
Contact: Svenska Elektriska Kommissionen Stockholm 5, Box 5177, S-102 44, Sweden
Vocabularies:
Subject(s): electrical engineering
UDC 621. 3
Language(s): Fr; En; Ru; De; Es; It; Nl; Pl; Sv

● *28* *(Commission/Committee)*
JCG/WGS^S "Radiolocation, direction finding, radar-radionavigation"
Parent organization: International Electrotechnical Commission
Contact: Comité électrotechnique français, Paris Cedex 16, 12, Place des Etats-Unis, F-75783, France
Vocabularies:
Subject(s): electrical engineering
UDC 621. 3
Language(s): Fr; En; Ru; De; Es; It; Nl; Pl; Sv

● *29* *(Commission/Committee)*
TC 1/WG 96 "Electrochemistry, electrometallurgy"
Parent organization: International Electrotechnical Commission
Contact: Comité électrotechnique français, Paris Cedex 16, 12, Place des Etats-Unis, F-75783, France
Vocabularies:
Subject(s): electrical engineering
UDC 621. 3
Language(s): Fr; En; Ru; De; Es; It; Nl; Pl; Sv

● *30* *(Commission/Committee)*
TC 1/WG 101 General concepts
Parent organization: International Electrotechnical Commission
Contact: Comité électrotechnique français, Paris Cedex 16, 12, Place des Etats-Unis, F-75783 France
Vocabularies:
Subject(s): electrical engineering ; electronics ; telecommunication
UDC 621. 3 ; UDC 654
Language(s): Fr; En; Ru; De; Es; It; Nl; Pl; Sv

● *31* *(Commission/Committee)*
TC 1/WG 601 "Generation, transmission and distribution of electric energy"
Parent organization: International Electrotechnical Commission
Contact: Verband Deutscher Elektrotechniker e. V.
Frankfurt/M. 70
Stresemannallee 21
D-6000, FRG
Vocabularies:
Subject(s): electrical engineering
UDC 621. 3
Language(s): Fr; En; Ru; De; Es; It; Nl; Pl; Sv

● *32* *(Commission/Committee)*
TC 1/WG 631 "Power plants for production of energy by nuclear means"
Parent organization: International Electrotechnical Commission
Contact: U. S. National Committee of the IEC
American National Standards Institute Inc.
New York
N. Y. 10018
U. S. A. 1430 Broadway
Vocabularies:
Subject(s): electrical engineering
UDC 621. 3
Language(s): Fr; En; Ru; De; Es; It; Nl; Pl; Sv

● *33* *(Commission/Committee)*
TC 1/WG 691 "Tariffs for electrical energy"
Parent organization: International Electrotechnical Commission
Contact: Dansk Elektroteknisk Komite
København K.
Strandgade 36, st.
DK-1401, Denmark
Vocabularies:
Subject(s): electrical engineering
UDC 621. 3
Language(s): Fr; En; Ru; De; Es; It; Nl; Pl; Sv

● *34* *(Commission/Committee)*
TC 1/WG 821 "Signalling and security apparatus for railways and roads"
Parent organization: International Electrotechnical Commission
Contact: Comité électrotechnique français, Paris Cedex 16, 12, Place des Etats-Unis, F-75783, France
Vocabularies:
Subject(s): electrical engineering
UDC 621. 3
Language(s): Fr; En; Ru; De; Es; It; Nl; Sv

● *35* *(Commission/Committee)*
TC 1/WG 871 "Electronic optics"
Parent organization: International Electrotechnical Commission
Contact: Vacant
Vocabularies:
Subject(s): electrical engineering
UDC 621. 3
Language(s): Fr; En; Ru; De; Es; It; Nl; Pl; Sv

● *36* *(Commission/Committee)*
TC 1/WG 881 "Radiology and radiological physics"

Parent organization: International Electrotechnical Commission
Contact: U. S. Committee of the IEC, American National Standards Institute Inc., New York, N. Y., 10018, USA
1430 Broadway,
Vocabularies:
Subject(s): electrical engineering
UDC 621. 3
Language(s): Fr; En; Ru; De; Es; It; Nl; Pl; Sv

● *37* *(Commission/Committee)*
TC 2/WG 2 "Rotating machines"
Parent organization: International Electrotechnical Commission
Contact: Comité électrotechnique français
Paris Cedex 16
12, Place des Etats-Unis
F-75783, France
Vocabularies:
Subject(s): electrical engineering
UDC 621. 3
Language(s): Fr; En; Ru; De; Es; It; Nl; Pl; Sv

● *38* *(Commission/Committee)*
TC 9/WG 4 "Electric traction"
Parent organization: International Electrotechnical Commission
Contact: Comité électrotechnique suisse, Association suisse des électriciens, Zürich, Seefeldstraße 301, Postfach, CH-8034, Switzerland
Vocabularies:
Subject(s): electrical engineering
UDC 621. 3
Language(s): Fr; En; Ru; De; Es; It; Nl; Pl; Sv

● *39* *(Commission/Committee)*
TC 10 "Liquid and gaseous dielectrics"
Parent organization: International Electrotechnical Commission
Contact: Comité électrotechnique belge
Bruxelles 3, Galerie Ravenstein
B-1000 Belgium
Vocabularies:
Subject(s): electrical engineering
UDC 621. 3
Language(s): Fr; En; Ru; De; Es; It; Nl; Pl; Sv

● *40* *(Commission/Committee)*
TC 11 "Overhead lines"
Parent organization: International Electrotechnical Commission
Contact: Comitato Elettrotecnico Italiano
Milano
Viale Monza, 259
I-20126, Italy
Vocabularies:
Subject(s): electrical engineering
UDC 621. 3
Language(s): Fr; En; Ru; De; Es; It; Nl; Pl; Sv

● *41* *(Commission/Committee)*
TC 13/WG 1 "Electrical measurements"
Parent organization: International Electrotechnical Commission
Contact: Magyar Szabvanyügyi Hivatal
Budapest 9
Ullói-ut 25 PF. 24
H-1450
Vocabularies:
Subject(s): electrical engineering
UDC 621. 3
Language(s): Fr; En; Ru; De; Es; It; Nl; Pl; Sv

● *42* *(Commission/Committee)*
TC 13/WG 1 "General terms on electrical and electronic measurement"
Parent organization: International Electrotechnical Commission
Contact: Magyar Szabvanyügyi Hivatal
Budapest 9
Ullói-ut 25 PF. 24 H-1150 Hungary
Vocabularies:
Subject(s): electrical engineering
UDC 621. 3
Language(s): Fr; En; Ru; De; Es; It; Nl; Pl; Sv

● *43* *(Commission/Committee)*
TC 14 "Transformers"
Parent organization: International Electrotechnical Commission
Contact: UK British Electrotechnical Committee
British Standards Institution
London 2, Park Street, W1A 2BS, U. K.
Vocabularies:
Subject(s): electrical engineering
UDC 621. 3
Language(s): Fr; En; Ru; De; Es; It; Nl; Pl; Sv

● *44* *(Commission/Committee)*
TC 15 "Solid insulating materials"
Parent organization: International Electrotechnical Commission
Contact: Comitato Elettrotecnico Italiano
Milano
Viale Monza, 259
I-20126, Italy
Vocabularies:
Subject(s): electrical engineering
UDC 621. 3
Language(s): Fr; En; Ru; De; Es; It; Nl; Pl; Sv

● *45* *(Commission/Committee)*
TC 17/WG 1 "Switchgear and controlgear"
Parent organization: International Electrotechnical Commission
Contact: Svenska Elektriska Kommissionen
Stockholm 5
Box 5177
S-102 44, Sweden
Vocabularies:
Subject(s): electrical engineering
UDC 621. 3
Language(s): Fr; En; Ru; De; Es; It; Nl; Pl; Sv

● *46* *(Commission/Committee)*
TC 20/WG 13 "Electric cables"
Parent organization: International Electrotechnical Commission
Contact: Comité électrotechnique français
Paris Cedex 16
12, Place des Etats-Unis
F-75783, France
Vocabularies:
Subject(s): electrical engineering
UDC 621. 3
Language(s): Fr; En; Ru; De; Es; It; Nl; Pl; Sv

● *47* *(Commission/Committee)*
TC 21 "Accumulators"
Parent organization: International Electrotechnical Commission
Contact: Úrad pro normalizaci a měřeni
Praha 1
Václavské nam, 19
CS-1134 47, Czechoslovakia
Vocabularies:
Subject(s): electrical engineering
UDC 621. 3
Language(s): Fr; En; Ru; De; Es; It; Nl; Pl; Sv

● *48* *(Commission/Committee)*
TC 22/WG 1 "Power electronics"
Parent organization: International Electrotechnical Commission

Contact: Nederlands Elektrotechnisch Comite
Nederlands Normalisatie-instituut
Delft
Kalfjeslaan 2
NL-2600 GB
Netherlands
Vocabularies:
Subject(s): electrical engineering
UDC 621. 3
Language(s): Fr; En; Ru; De; Es; It; Nl; Pl; Sv

● ***49*** *(Commission/Committee)*
TC 22/WG 1 "Transductors"
Parent organization: International Electrotechnical Commission
Contact: Nederlands Elektrotechnisch Comite
Nederlands Normalisatie-instituut
Delft
Kalfjeslaan 2
NL-2600 GB
Netherlands
Vocabularies:
Subject(s): electrical engineering
UDC 621. 3
Language(s): Fr; En; Ru; De; Es; It; Nl; Pl; Sv

● ***50*** *(Commission/Committee)*
TC 23/WG 2 "Electrical accessories"
Parent organization: International Electrotechnical Commission
Contact: Comité électrotechnique belge
Bruxelles
3, Galerie Ravenstein
B-1000 Belgium
Vocabularies:
Subject(s): electrical engineering
UDC 621. 3
Language(s): Fr; En; Ru; De; Es; It; Nl; Pl; Sv

● ***51*** *(Commission/Committee)*
TC 26 "Electric welding"
Parent organization: International Electrotechnical Commission
Contact: Comité électrotechnique belge, Bruxelles, 3, Galerie Ravenstein, B-1000, Belgium
Vocabularies:
Subject(s): electrical engineering
UDC 621. 3
Language(s): Fr; En; Ru; De; Es; It; Nl; Pl; Sv

● ***52*** *(Commission/Committee)*
TC 27 "Industrial electroheating"
Parent organization: International Electrotechnical Commission
Contact: Polski Komitet Normalizacji i Miar, Ul. Elektoralna 2, Warszawa 5I, Poland
Vocabularies:
Subject(s): electrical engineering
UDC 621. 3
Language(s): Fr; En; Ru; De; Es; It; Nl; Pl; Sv

● ***53*** *(Commission/Committee)*
TC 28 "Insulation coordination"
Parent organization: International Electrotechnical Commission
Contact: Comité électrotechnique français
Paris Cedex 16
12, Place des Etats-Unis
F-75783, France
Vocabularies:
Subject(s): electrical engineering
UDC 621. 3
Language(s): Fr; En; Ru; De; Es; It; Nl; Pl; Sv

● ***54*** *(Commission/Committee)*
TC 29/WG 15 "Electro-acoustics"
Parent organization: International Electrotechnical Commission
Contact: U. S. National Committee of the IEC, American National Standards Institute Inc., New York, N. Y. 10018, USA
1430 Broadway,
Vocabularies:
Subject(s): electrical engineering
UDC 621. 3
Language(s): Fr; En; Ru; De; Es; It; Nl; Pl; Sv

● ***55*** *(Commission/Committee)*
TC 31 "Flameproof equipments"
Parent organization: International Electrotechnical Commission
Contact: British Electrotechnical Committee
British Standards Institution
London
2, Park Street, W1A 2BS, U. K.
Vocabularies:
Subject(s): electrical engineering
UDC 621. 3
Language(s): Fr; En; Ru; De; Es; It; Nl; Pl; Sv

● ***56*** *(Commission/Committee)*
TC 32/WG 1 "Fuses"
Parent organization: International Electrotechnical Commission
Contact: Comité électrotechnique français
Paris Cedex 16
12, Place des Etats-Unis
F-75783, France
Vocabularies:
Subject(s): electrical engineering
UDC 621. 3
Language(s): Fr; En; Ru; De; Es; It; Nl; Pl; Sv

● ***57*** *(Commission/Committee)*
TC 33 "Power capacitors"
Parent organization: International Electrotechnical Commission
Contact: Nederlands Elektrotechnisch Comite
Nederlands Normalisatie-instituut
Delft
Kalfjeslaan 2
NL-2600 GB
Netherlands
Vocabularies:
Subject(s): electrical engineering
UDC 621. 3
Language(s): Fr; En; Ru; De; Es; It; Nl; Pl; Sv

● ***58*** *(Commission/Committee)*
TC 34 "Lighting"
Parent organization: International Electrotechnical Commission
Contact: Comité électrotechnique français, Paris Cedex 16, 12, Place des Etats-Unis, F-75783, France
Vocabularies:
Subject(s): electrical engineering
UDC 621. 3
Language(s): Fr; En; Ru; De; Es; It; Nl; Pl; Sv

● ***59*** *(Commission/Committee)*
TC 35 "Primary cells and batteries"
Parent organization: International Electrotechnical Commission
Contact: Comité électrotechnique français
Paris Cedex 16
12, Place des Etats-Unis
F-75783, France
Vocabularies:
Subject(s): electrical engineering
UDC 621. 3
Language(s): Fr; En; Ru; De; Es; It; Nl; Pl; Sv

● ***60*** *(Commission/Committee)*
TC 36 "Insulators"

Parent organization: International Electrotechnical Commission
Contact: Comitato Elettrotecnico Italiano
Milano
Viale Monza, 259
I-20126, Italy
Vocabularies:
Subject(s): electrical engineering
UDC 621. 3
Language(s): Fr; En; Ru; De; Es; It; Nl; Pl; Sv

● ***61*** *(Commission/Committee)*
TC 37 "Lightning arresters"
Parent organization: International Electrotechnical Commission
Contact: U. S. National Committee of the IEC
American National Standards Institute Inc.
New York
N. Y. 10018
U. S. A.
1430 Broadway
Vocabularies:
Subject(s): electrical engineering
UDC 621. 3
Language(s): Fr; En; Ru; De; Es; It; Nl; Pl; Sv

● ***62*** *(Commission/Committee)*
TC 38 "Instruments transformers"
Parent organization: International Electrotechnical Commission
Contact: Verband Deutscher Elektrotechniker e. V.
Frankfurt/M. 70
Stresemannallee 21
D-6000 FRG
Vocabularies:
Subject(s): electrical engineering
UDC 621. 3
Language(s): Fr; En; Ru; De; Es; It; Nl; Pl; Sv

● ***63*** *(Commission/Committee)*
TC 39/WG 1 "Electronic tubes"
Parent organization: International Electrotechnical Commission
Contact: Comité électrotechnique français
Paris Cedex 16
12, Place des Etats-Unis
F-75783, France
Vocabularies:
Subject(s): electrical engineering
UDC 621. 3
Language(s): Fr; En; Ru; De; Es; It; Nl; Pl; Sv

● ***64*** *(Commission/Committee)*
TC 40/WG 6 "Capacitors and resistors for electronic equipment"
Parent organization: International Electrotechnical Commission
Contact: Nederlands Elektrotechnisch Comite
Nederlands Normalisatie-instituut
Delft
Kalfjeslaan 2
NL-2600 GB
Netherlands
Vocabularies:
Subject(s): electrical engineering
UDC 621. 3
Language(s): Fr; En; Ru; De; Es; It; Nl; Pl; Sv

● ***65*** *(Commission/Committee)*
TC 41/WG 1 "Electrical relays"
Parent organization: International Electrotechnical Commission
Contact: Comité électrotechnique français
Paris Cedex 16
12, Place des Etats-Unis
F-75783, France
Vocabularies:
Subject(s): electrical engineering
UDC 621. 3
Language(s): Fr; En; Ru; De; Es; It; Nl; Pl; Sv

● ***66*** *(Commission/Committee)*
TC 45/WG 1 "Electrical detection and measurement of ionizing radiations"
Parent organization: International Electrotechnical Commission
Contact: Comité électrotechnique français
Paris Cedex 16
12, Place des Etats-Unis, F-75783, France
Vocabularies:
Subject(s): electrical engineering
UDC 621. 3
Language(s): Fr; En; Ru; De; Es; It; Nl; Pl; Sv

● ***67*** *(Commission/Committee)*
TC 47/WG 1 "Semiconductor devices and integrated circuits"
Parent organization: International Electrotechnical Commission
Contact: British Electrotechnical Committee
British Standards Institution
London
2, Park Street, W1A 2BS, U. K.
Vocabularies:
Subject(s): electrical engineering
UDC 621. 3
Language(s): Fr; En; Ru; De; Es; It; Nl; Pl; Sv

● ***68*** *(Commission/Committee)*
TC 48 "Electromechanical components for electronic equipment"
Parent organization: International Electrotechnical Commission
Contact: Vacant
Vocabularies:
Subject(s): electrical engineering
UDC 621. 3
Language(s): Fr; En; Ru; De; Es; It; Nl; Pl; Sv

● ***69*** *(Commission/Committee)*
TC 48/WG 1 "Electromechanical components for electronic equipment"
Parent organization: International Electrotechnical Commission
Contact: U. S. National Committee of the IEC
American National Standards Institute Inc. New York
N. Y. 10018
U. S. A.
1430 Broadway
Vocabularies:
Subject(s): electrical engineering
UDC 621. 3
Language(s): Fr; En; Ru; De; Es; It; Nl; Pl; Sv

● ***70*** *(Commission/Committee)*
TC 49 "Piezoelectric crystals"
Parent organization: International Electrotechnical Commission
Contact: Gosudarstvennyj komitet standartov
Soveta ministrov SSSR
Moskva 49
Leninsky pr. 9
SU-117049, USSR
Vocabularies:
Subject(s): electrical engineering
UDC 621. 3
Language(s): Fr; En; Ru; De; Es; It; Nl; Pl; Sv

● ***71*** *(Commission/Committee)*
TC 49 "Piezoelectric devices for frequency control and selection"
Parent organization: International Electrotechnical Commission

Contact: Gosudarstvennyj komitet standartov
Soveta ministrov SSSR
Moskva M-49
Leninsky pr. 9
SU-117049, USSR
Vocabularies:
Subject(s): electrical engineering
UDC 621. 3
Language(s): Fr; En; Ru; De; Es; It; Nl; Pl; Sv

● ***72*** *(Commission/Committee)*
TC 51/WG 3 "Magnetic components and ferrite materials"
Parent organization: International Electrotechnical Commission
Contact: Nederlands Elektrotechnisch Comite
Nederlands Normalisatie-instituut
Delft
Kalfjeslaan 2
NL-2600 GB
Netherlands
Vocabularies:
Subject(s): electrical engineering
UDC 621. 3
Language(s): Fr; En; Ru; De; Es; It; Nl; Pl; Sv

● ***73*** *(Commission/Committee)*
TC 51/WG 3 "Magnetic materials"
Parent organization: International Electrotechnical Commission
Contact: Nederlands Elektrotechnisch Comite
Nederlands Normalisatie-Instituut
Delft
Kalfjeslaan 2
NL-2600 GB, Netherlands
Vocabularies:
Subject(s): electrical engineering
UDC 621. 3
Language(s): Fr; En; Ru; De; Es; It; Nl; Pl; Sv

● ***74*** *(Commission/Committee)*
TC 51/WG 3 "Vocabulary on magnetism"
Parent organization: International Electrotechnical Commission
Contact: Nederlands Elektrotechnisch Comite, Nederlands Normalisatie-instituut
NL-2600 GB Delft, Kalfjeslaan 2, NL-2600 GB, Netherlands
Vocabularies:
Subject(s): electrical engineering
UDC 621. 3
Language(s): Fr; En; Ru; De; Es; It; Nl; Pl; Sv

● ***75*** *(Commission/Committee)*
TC 52/WG 5 "Printed circuits"
Parent organization: International Electrotechnical Commission
Contact: Nederlands Elektrotechnisch Comite
Nederlands Normalisatie-instituut
Delft
Kalfjeslaan 2
NL-2600 GB
Netherlands
Vocabularies:
Subject(s): electrical engineering
UDC 621. 3
Language(s): Fr; En; Ru; De; Es; It; Nl; Pl; Sv

● ***76*** *(Commission/Committee)*
TC 55 "Winding wires"
Parent organization: International Electrotechnical Commission
Contact: Verband Deutscher Elektrotechniker e. V.
Frankfurt/M. 70
Stresemannallee 21, FRG
Vocabularies:
Subject(s): electrical engineering
UDC 621. 3
Language(s): Fr; En; Ru; De; Es; It; Nl; Pl; Sv

● ***77*** *(Commission/Committee)*
TC 56/WG 1 "Reliability"
Parent organization: International Electrotechnical Commission
Contact: British Electrotechnical Committee, British Standards Institution London,
2, Park Street, W1A 2BS, U. K.
Vocabularies:
Subject(s): electrical engineering
UDC 621. 3
Language(s): Fr; En; Ru; De; Es; It; Nl; Pl; Sv

● ***78*** *(Commission/Committee)*
TC 57 "Telecontrol"
Parent organization: International Electrotechnical Commission
Contact: Vacant
Vocabularies:
Subject(s): electrical engineering
UDC 621. 3
Language(s): Fr; En; Ru; De; Es; It; Nl; Pl; Sv

● ***79*** *(Commission/Committee)*
TC 59 "Household electrical applications"
Parent organization: International Electrotechnical Commissions
Contact: Comité électrotechnique français, Paris Cedex 16, 12, place des Etats-Unis, F-75783, France
Vocabularies:
Subject(s): electrical engineering
UDC 621. 3
Language(s): Fr; En; Ru; De; Es; It; Nl; Pl; Sv

● ***80*** *(Commission/Committee)*
TC 60 "Educational and training equipment"
Parent organization: International Electrotechnical Commission
Contact: Nederlands Elektrotechnisch Comite
Nederlands Normalisatie-instituut
Delft
Kalfjeslaan 2
NL-2600 GB
Netherlands
Vocabularies:
Subject(s): electrical engineering
UDC 621. 3
Language(s): Fr; En; Ru; De; Es; It; Nl; Pl; Sv

● ***81*** *(Commission/Committee)*
TC 60/WG 1 "Recording and reproducing of sound and video"
Parent organization: International Electrotechnical Commission
Contact: Comité électrotechnique français, Paris Cedex 16, 12, Place des Etats-Unis, F-75783, France
Vocabularies:
Subject(s): electrical engineering
UDC 621. 3
Language(s): Fr; En; Ru; De; Es; It; Nl; Pl; Sv

● ***82*** *(Commission/Committee)*
TC 62/SC 62A/4 "Electrobiology"
Parent organization: International Electrotechnical Commission
Contact: Dansk Elektroteknisk Komite, København K., Strandgade 36, st., DK-1401, Denmark
Vocabularies:
Subject(s): electrical engineering
UDC 621. 3
Language(s): Fr; En; Ru; De; Es; It; Nl; Pl; Sv

● ***83*** *(Commission/Committee)*
TC 64 "Electrical installations in buildings"

Parent organization: International Electrotechnical Commission
Contact: Vacant
Vocabularies:
Subject(s): electrical engineering
UDC 621. 3
Language(s): Fr; En; Ru; De; Es; It; Nl; Pl; Sv

● ***84*** *(Commission/Committee)*

TC 65/WG 1 "Automatic control and regulation, servo-mechanisms"

Parent organization: International Electrotechnical Commission
Contact: Comité électrotechnique belge
Bruxelles
3, Galerie Ravenstein
B-1000 Belgium
Vocabularies:
Subject(s): electrical engineering
UDC 621. 3
Language(s): Fr; En; Ru; De; Es; It; Nl; Pl; Sv

● ***85*** *(Commission/Committee)*

TC CISPR/WG 9 "Radio interference"

Parent organization: International Electrotechnical Commission
Contact: British Electrotechnical Committee, British Standards Institution, London,
2, Park Street, W1A 2BS, U. K.
Vocabularies:
Subject(s): electrical engineering
UDC 621. 3
Language(s): Fr; En; Ru; De; Es; It; Nl; Pl; Sv

● ***86*** *(Commission/Committee)*

Commission internationale de l'éclairage, Technical Committee 1. 1 "Terminology"

[International Commission on Illumination (not official); Internationale Beleuchtungskommission (De)]

CIE

52, Boulevard Malesherbes — Phone: 387 21 21
Paris, F-75008 — Telegr.: COMECLAIR-PARIS
France
Chairperson: Dr. J. Terrien, Chairman of Technical Committee 1. 1 "Terminology"
Established: 1913
Vocabularies:
Subject(s): light; lighting
UDC 628. 9
Language(s): Fr; En; De
Vocabulary standards:
Subject(s): light; lighting
UDC 628. 9
Language(s): En; Fr; De; Ru
Publications: *International Lighting Vocabulary,* CIE-Publication No. 17

● ***87*** *(Commission/Committee)*

Commission on Symbols, Units, Nomenclature, Atomic Masses and Fundamental Constants "SUN-AMCO"

Parent organization: International Union of Pure and Applied Physics
Université Laval — Phone: (418) 656–2131
Québec 10
Canada
Contact: Prof. Robert C. Barber
Secretary
IUPAP SUN-AMCO Commission
Department of Physics
University of Manitoba
Winnipeg
Canada R3T 2N2
Committee members: 10 members; 4 associate members
Chairperson: Mr. Leonardo Villena
Vice-chairperson: Mr. Robert C. Barber, Secretary
Established: 1979 - "SUN-AMCO" formed by amalgamation of "SUN" Commission (formed in 1931) and Commission on Atomic Masses and Fundamental Constants (formed in 1960)
Vocabularies:
Subject(s): physics
UDC 53
Language(s): En; Fr
Publications: IUPAP/S. U. N. COMMISSION. *Symbols, units and nomenclature in physics.* Published with the financial support of Unesco. Document U. I. P. 20 (1978)

● ***88*** *(Commission/Committee)*

Committee on Conceptual and Terminological Analysis

COCTA

Parent organization: International Social Science Council (ISSC); International Political Science Association; International Sociological Association
1, rue Miollis — Phone: 577–16–10
Paris, F-75015 — Telegr.: SOSCICOUN PARIS
ISSC, Unesco, France
Contact: Prof. Fred W. Riggs, Dept. of Political Science, University of Hawaii, 2424 Maile Way, Honolulu, Hawaii 96822, USA
Committee members: 150
Chairperson: Prof. Fred W. Riggs
Vice-chairperson: Prof. Henry Teune; Prof. George J. Graham, Jr.
Established: 1970
Vocabularies: metalanguage (concepts and terms) used in conceptual and terminological analysis in the social sciences
Subject(s): social sciences
UDC 3
Language(s): En
Vocabulary standards: COCTA supports descriptive studies of actual usage and preparation of glossaries; it does not prescribe any standardized vocabularies
Theory: Theory and design of glossaries (descriptive) for use in social science subject fields. Also detailed studies of the meanings of key terms as they are used by social scientists
Publications: *COCTA News;* published thrice a year in *International Classification;* also available separately for US $5 a year; Memoranda and papers by COCTA members are available for private distribution and comment as announced in *COCTA News*; a symposium, *Key Concepts in the Social Sciences* is in preparation, edited by Prof. G. Sartori; *The Tower of Babel* (1975) by G. Sartori, F. W. Riggs, H. Teune (out of print); *Conference on Conceptual and Terminological Analysis in the Social Sciences* (CONTA), held in Bielefeld (FRG), May 24–27, 1981, sponsored by COCTA, in cooperation with the German UNESCO Commission; FID/CR+ SD; Informationszentrum Sozialwissenschaften, Bonn; Deutsche Gesellschaft für Dokumentation, Frankfurt; and Gesellschaft für Klassifikation e. V., Frankfurt. Proceedings will be published. Abstracts of 33 papers available as a preconference document, from Dr. I. Dahlberg, Woogstr. 36a, D-6000 Frankfurt/Main 50, FRG

● ***89*** *(Commission/Committee)*

Committee on Storage, Automatic Processing and Retrieval of Geological Data

COGEODATA

Parent organization: International Union of Geological Sciences
601 Booth Street — Phone: 613–995–4927
Ottawa, Ontario, K1A OE8 — Telex: 0533117 EMAR-OTT
Canada
Contact: Dr. R. Sinding-Larsen
P. O. Box 3006
7001 Trondheim/Norway
Committee members: 12
Chairperson: Dr. R. Sinding-Larsen
Vice-chairperson: Dr. J. Hruska

Established: 1972
Vocabulary standards:
Subject(s): geology; computer application
UDC 541. 1/. 4; UDC 681. 3
Publications: COGEODATA Newsletter; without charge

● ***90*** *(Organization)*

Council for International Organizations of Medical Sciences/Conseil des Organisations internationales des sciences médicales
CIOMS/COISM
Parent organization: World Health Organization and Unesco
c/o WHO, Avenue Appia — Phone: 022/91 34 06
Genéva, CH-1211 — Telex: 27 821
Switzerland
Staff:
Head: Mr. Zbigniew Bankowski, M. D., Executive Secretary
Expert(s): 4
Staff categories:
2 terminologist(s) (theory); 1 translator(s)
Founded: 1949
Documentation:
bibliographic data
factographic data
Subject(s): diseases (medicine)
UDC 61
Language(s): En; Fr; Es; De; (and others)
Vocabularies:
Subject(s): medicine
UDC 61
Language(s): En; Fr; De; Es
Vocabulary standards:
Subject(s): medicine
UDC 61
Language(s): En; Fr; De; Es
Publications: *International Nomenclature of Diseases. Diseases of the Lower Respiratory Tract.* Geneva: CIOMS, 1979, 128 p., A5. Others in preparation, i. e. infectious diseases

(Project(s))
Responsible: CIOMS/WHO Technical Steering Committee
Title: "International Nomenclature of Diseases"
Project type: institutional; in progress
Principal investigator: Dr. Zbigniew Bankowski, CIOMS; Dr. K. Kupka, WHO
Collaborators: 7
Scope and/or subject matter: Objective of the IND is to privide, for every morbid entity, a single recommended name
Method of work: Consultation of existing recommendations; draft is distributed to a large number of individual experts and professional associations
Source material: Existing recommendations
Publications: Nomenclature on infectious diseases and neurological diseases (forthcoming)

● ***91*** *(Organization)*

Division of the Unesco Library, Archives and Documentation Services/Division de la bibliothèque de l'Unesco, des archives et des services de documentation/División de la biblioteca de la Unesco, de los archivos y de los servicios de documentación
LAD
Parent organization: UNESCO
7, Place de Fontenoy — Phone: 755-16-10
Paris, F-75007 — Telex: 27602 PARIS
France — Telegr.: UNESCO PARIS
Head: Mr. M. Pobukovsky
Expert(s): Ms. J. Forestier, chief of section
Founded: 1972
Documentation: bibliographic data
Subject(s): science; technology; education; social sciences; culture humanities; communication; information, libraries; archives
UDC 3/6
Language(s): En (and others)
Publications: Aitchinson, J. *Unesco Thesaurus*: a structured list of descriptors for indexing and retrieving literature in the fields of education, science, social science, culture and communication. Paris: Unesco, 1977, 2 vol.: 485 p. + 530 p. ISBN 92-3-101469-2. To be published in Fr in 1981 and in Es in 1982.
Services: reference; loan; reproduction; referral
No service charges
Operation:
computer-assisted
Data base description:
Name: CDS (Computerized Documentation System); CDTHES (Unesco Thesaurus) ; CROSS-REF (External Cross Reference and Translations)
Size: CDS (40 000); CDTHES (12 000) ; CROSSREF (8 000)
Language(s): En
Classification: UNESCO thesaurus; UDC
Terminological data:
definitions; explanations; equivalents; other data elements
Data processing:
Hardware: IBM; 3031; 2 MB; printers; terminals (Memorex 1377 etc.)
Software: operating system: OS/VSI; operation mode: batch; real time; programming language(s): ASM; FORTRAN; COBOL; PL1; APL; data base management system/retrieval documentation system: CDS/ISIS/CDS/ISIS
Access:
on-line; off-line
Data available:
free of charge

● ***92*** *(Commission/Committee)*

Documentation and Astronomical Numerical Data (Commission 5)
Parent organization: International Astronomical Union/Union astronomique internationale
61, avenue de l'Observatoire — Phone: (1) 325 83 58
Paris, F-75016 — Telegr.: UNASTRINTER Paris
France
Contact: Mr. W. D. Heintz
Dept. of Astronomy
Swarthmore College
Swarthmore
PA. 19081
USA;
Contact: Work relating to CODATA: B. Hauck, Dept. d'Astronomie, Université de Lausanne, CH-1290 Chavannes-des-Bois, Switzerland
Committee members: 30
Chairperson: Mr. Wulff D. Heintz
Vice-chairperson: Mr. George A. Wilkins
Established: 1921
Vocabularies:
Subject(s): astronomy; astrophysics
UDC 52
Language(s): En (is basic language for key-word lists)
Publications: Commission Reports issued in the Transactions of the International Astronomical Union (Reidel Co., Dordrecht, Netherlands) ; all other publications are published by the individual (bibliographical or numerical) data centers

(Project(s))
Title: At present projects are not executed by this Commission but by the bibliographical centers, although partly under the auspices, and/or with support of this Commission

● ***93*** *(Organization)*

Food and Agriculture Organization of the United

Nations. Terminology and Reference Section/ Organisation des Nations Unies pour l'alimentation et l'agriculture. Section terminologie et références/ Organización de las Naciones Unidas para la agricultura y la alimentación. Sección de terminología y referencias (Es)
FAO-TERM
Parent organization: United Nations
Via delle terme di Caracalla — Phone: 57971
Roma, I-00152 — Telex: 610181 FAO I
Italy — Telegr.: FOODAGRI ROME
Staff: 9
Head: Dr. P. Savary (Publications Division)
Expert(s): Mrs. Giuliana Grütter-Roviglioni
Staff categories:
2 documentalist(s)/librarian(s); 1 other professional (s)
Founded: 1965
Documentation:
bibliographic data
factographic data
Subject(s): agriculture; nutrition; economics; statistics; forestry; meteorology; plants; irrigation; commodities; rural development; food; fisheries
UDC 63; UDC 612; UDC 33; UDC 311; UDC 634; UDC 551. 5; UDC 631; UDC 626; UDC 338; UDC 711. 3; UDC 612. 39; UDC 639. 2
Language(s): En; Fr; Es; Ar; Ch
Vocabularies:
Subject(s): food; agriculture
UDC 63; UDC 612
Language(s): En; Fr; Es; Ar; Ch
Publications: Terminology Bulletins such as: *Selected terms in nutrition and home economics*. 1968, 87 p., in En, Fr, Es, De; *Selected terms in the field of agricultural development aid*. 1969, 32 p., in En, Fr, Es; *Selected terms in fish culture*. 1972, 106 p., in En, Fr, Es, La; *Names of countries and their capital cities*, including adjectives of nationality and currency units. 1980, 46 p., in En, Fr, Es; *Vocabulary of agricultural and other terms in use by FAO*. 1969, 625 p., in En, Ar;
Terms on environment related to agriculture. 1979, 170 p., in En, Fr, Es, Ar; *Administrative and financial terms, including those related to FAO field programme operations*. 1973, 63 p., in En, Fr, Es; *Selected terms on bananas*. 1979, 63 p., in En, Fr, Es, De, La, Ar; *Plants and plant products of economic importance*. 1973, 328 p., in En, Fr, Es, De, La; *Rice terminology*. 1979, 108 p., in En, Fr, Es, Ar; *Plywood, fibreboard and particle board*. 1980, 201 p., in En, Fr, Es, De, It, Ar; *Milk and milk products: terminology*. 1980, 120 p., in En, Fr, Es, La, Ar; *Agricultural credit*. 1978, 218 p., in En, Fr, Es; and many others.
Terminology Notes such as: *Selected terms in data processing*. 1970, 37 p., in En, Fr; *Selected terms used in fish culture*. 1971, 26 p., in En, Es; French equivalents of selected English technical terms as published in the *Journal officiel de la République française*. 1974, 74 p., in En, Fr; *Dictionaries and vocabularies in the field of forestry*. 1972, 38 p.; *Dictionaries and encyclopaedias in the field of fisheries*. 1973, 20 p.; *Dictionaries and vocabularies in the Terminology and Reference Library*. 1966–77, 5th ed., 1978, 170 p.
Terminology Bulletins planned or in preparation: Fish culture (in preparation the Arabic version); Administration and financial terms (in preparation); Plants and plant products (in preparation the Arabic version); Food and nutrition (in preparation the Arabic version); Agricultural credit (in preparation the Arabic version); Remote sensing (with definitions in English) (in preparation); Forestry (planned); Tobacco (in preparation); AGRIS terminology (in preparation); Meat (planned)
Services: reference; loan; translation
No service charges
Operation:
manual
Access: via telephone / via telex
Data available:
on exchange basis; free of charge

● ***94*** *(Commission/Committee)*
IGU-Commission International Geographical Terminology (since 1980: IGU-Working group)
[UGI-Commission de terminologie géographique internationale; IGU-Kommission Internationale Geographische Terminologie (De)]
Parent organization: International Geographical Union (IGU)
Werderring 4
Freiburg i. Brsg., D-7800
Contact: Prof. Dr. E. Meynen
Langenbergweg 82
D-5300 Bonn 2 (Bad Godesberg)
FRG, Tel. 0228/34 29 82
Committee members: 10 effective members, 25 corresponding members
Chairperson: Prof. E. M. Yates, London (since 1980)
Vice-chairperson: Prof. Dr. E. Meynen (Secretary since 1980)
Established: 1968 (1980 reestablished as IGU-Working group)
Vocabularies:
Subject(s): geography (physical as well as human geography)
UDC 91
Language(s): En; Fr; De; It; Ja; Ru; Es
Vocabulary standards:
Subject(s): geography (physical as well as human geography)
UDC 91
Language(s): De (source language) ; En; Fr; It; Ja; Ru; Es
Publications: 1) *International Dictionary of Technical Terms used in Geography*, Pilot-sample, submitted on the occasion of the 22nd International Geographical Congress, Montré al, August 1972. Franz Steiner Verlag, Wiesbaden 1972; 2) *International Geographical Terminology, Sixlingual Synopsis of the 2. 400 Keyterms* (Proof copy), submitted on the occasion of the 24th International Geographical Congress, Tokyo, 1980. Franz Steiner Verlag, Wiesbaden, 1980

(Project(s))
Responsible: International Geographical Terminology
Title: International Geographical Glossary [Internationales Geographisches Glossarium (De)]
Project type: commissioned project; in progress
Principal investigator: Prof. Dr. E. Meynen, Bonn, Chairman 1968–80, Secretary since 1980; Prof. E. M. Yates, London, Chairman since 1980
Collaborators: For De: 50; other national working groups are composed of different number of persons
Scope and/or subject matter: Geography, physical as well as human geography
Method of work: Teamwork of international and national working groups
Source material: *Bibliography of Mono- and Multilingual Dictionaries and Glossaries of Technical Terms used in Geography as well as in Related Natural and Social Sciences*. Wiesbaden: Franz Steiner Verlag, 1974, 246 p.
Publications: *International Geographical Glossary*; Published papers available on project

● ***95*** *(Organization)*
Institut international du froid/International Institute of Refrigeration
IIF/IIR
177, Boulevard Malesherbes — Phone: 227 32 35
Paris, F-75017
France
Staff: 10 (8 full time, 2 part time)
Head: Dr. A. Gac, Director
Staff categories:
2 documentalist(s)/librarian(s); 1 translator(s)
Founded: 1908

Vocabularies:
Subject(s): refrigeration; cryology; air-conditioning
UDC 621. 56/. 59
Language(s): En; Fr; Ru; De; Es; It; No
Publications: *New International Dictionary of Refrigeration*, in En, Fr, Ru, De, Es, It, No. Paris, 1978, 560p. This dictionary consists of 2 parts: the first part includes all entries in systematic order; the second part includes a series of alphabetical indexes in 7 languages, referring to the first part via reference numbers.
Services: translation
Service charges

● ***96*** *(Organization)*
Inter-Organisation Board for Information Systems/ Bureau interorganisations pour les systèmes d'information/Junta interorganizacional para sistemas de información (Es)
IOB/JISI(es)
Parent organization: United Nations
Palais des Nations
Genève, CH-1211
Switzerland
Phone: (022) 34 60 11 or 31 02 11
Telex: 28 96 96
Telegr.: UNATIONS GENEVA
Staff: 7
Head: Mme. Nathalie Dusoulier, Officer-in-Charge
Staff categories:
3 information specialist(s)
Vocabulary standards:
Subject(s): all fields (of interest to the United Nations)
UDC 0/9
Language(s): En; Fr; Es
Publications: 1. *Directory of United Nations Information Systems*, 2nd edition, 2 vols., 1980, En, Fr, Es, Ch. (United Nations sales publication); 2. *Broad Terms for United Nations Programmes and Activities*, 1979, En only. (United Nations sales publication); 3. *Bibliography of United Nations Thesauri, Classifications, Nomenclatures*, 1979. En only. (Free from IOB)
User aids: IOB Newsletter (restricted to staff members of the United Nations family of organizations)

(Project(s))
Title: Broad Terms for United Nations Programmes and Activities
Principal investigator: Mme. Nathalie Dusoulier
Scope and/or subject matter: All fields of activity of United Nations Organizations
Method of work: 28 organizations of the United Nations family have contributed terms describing their programmes and activities. The IOB is responsible for compilation of these terms.
Source material: Thesauri and other linguistic tools of the participating organizations. (Listed in "Bibliography of United Nations Thesauri, Classifications, Nomenclatures", issued by IOB, 1979. Available free of charge)
Publications: *Broad Terms for United Nations Programmes and Activities*, 1979, En only. (United Nations sales publication)

● ***97*** *(Organization)*
International Association of Dredging Companies
[Association internationale d'entreprises de dragage; Internationale Vereniging van Baggerondernemingen (Nl)]
I. A. D. C.; A. I. E. D.
Duinweg 21
P. O. Box 80521
's-Gravenhage, 2508 GM
Netherlands
Phone: 070 - 54 56 02
Telex: 31102 dune nl
Telegr.: Dredging, The Hague
Visiting address: Duinweg 21, The Hague
Staff: 10
Head: Dr. J. G. Th. Linssen
Expert(s): Mr. N. Oosterbaan
Staff categories:
1 information specialist(s); 1 other professional(s)
Founded: 1965
Documentation:
bibliographic data
factographic data
Subject(s): dredging
UDC 627. 5/. 7; UDC 621. 879. 447; UDC 626
Language(s): En; Fr; Nl
Vocabularies:
Subject(s): dredging
UDC 627. 5/. 7
Language(s): En; Fr; Nl
Vocabulary standards:
Subject(s): dredging
UDC 627. 5/. 7
Language(s): En; Fr; Nl
Publications: *International Dredging Abstracts*, ISSN 0308-1400; *Terra et Aqua* (English edition), ISSN 0376-6411; *Terra et Aqua* (French edition), ISSN 0376-7434; *IADC Newsletter*
Services: reference; loan; reproduction (limited); referral
Service charges
Operation:
manual
computer-assisted
computerized
Data base description:
Name: Dredging vocabulary
Size: 1500 entries
Language(s): En; Fr; Nl
Classification: Alphabetic
Terminological data:
equivalents
Access:
off-line
Data available:
with charge

(Project(s))
Title: Dredging dictionary (multilingual)
Principal investigator: Dr. J. G. Th. Linssen
Scope and/or subject matter: Dredging

Title: Dredging thesaurus
Principal investigator: N. Oosterbaan
Scope and/or subject matter: Dredging
Source material: Delft Hydraulics code; PIANC *Illustrated technical dictionary;* and others

Title: International Dredging Abstracts
Principal investigator: N. Oosterbaan
Collaborators: 3
Scope and/or subject matter: Dredging
Source material: 500 learned journals and trade journals
Publications: *International Dredging Abstracts* (ISSN 0308-1400)

● ***98*** *(Organization)*
International Bank for Reconstruction and Development (World Bank) Language Services Division/ Banque Internationale pour la reconstruction et le développement (Banque Mondiale) Division des services linguistiques/Banco internacional de reconstrucción y fomento (Banco Mondial) Division de los servicios linguisticos
IBRD/BIRD/BIRF
1818 H Street, N. W.
Washington, D. C., 20433
USA
Phone: (202) 676 0285
Telegr.: INTBAFRAD
Visiting address: 1990 K Street, N. W., Washington, D. C. 20433, USA
Staff: 2
Head: Mr. Robin F. Wollatt
Expert(s): Mr. John Alvey
Staff categories:
1 terminologist(s) (special field); 1 other professional(s)

Founded: 1946
Documentation:
bibliographic data
factographic data
Subject(s): finance; economics; science; technology
UDC 330; UDC 336; UDC 6
Language(s): En; Fr; Es; Pt; De; It
Vocabularies:
Subject(s): finance; economics; science; technology
UDC 330; UDC 336; UDC 6
Language(s): En; Fr; Es
Vocabulary standards:
Subject(s): finance
UDC 330
Language(s): En; Fr; Es
Publications: *World Bank Glossary* in En, Fr, Es; special ad hoc glossaries
Services: Provided by main division (normally available to staff only); reference; loan; reproduction; translation; referral
User aids: International Terminology Bulletin
Operation:
computer-assisted
Data base description:
Name: Cardex
Size: Small
Language(s): En; Fr; Es
Terminological data:
definitions; explanations; context; equivalents; other data elements
Access:
via telephone
Data available:
on exchange basis; free of charge

● ***99*** *(Commission/Committee)*
International Cartographical Association-Commission Multilingual Dictionary
[ICA-Commission Dictionnaire multilingue de termes techniques cartographiques; ICA-Kommission Mehrsprachiges Wörterbuch kartographischer Fachbegriffe (De)]
Parent organization: International Cartographical Association (ICA)
Meijerinksweg 9
Lonneker, NL-1832
The Netherlands
Contact: Prof. Dr. E. Meynen
Langenbergweg 82
D-5300 Bonn 2 (Bad Godesberg)
FRG, Tel. 0228/34 29 82
Chairperson: Prof. Dr. E. Meynen
Vice-chairperson: Prof. Dr. J. Neumann, D-7500 Karlsruhe, FRG
Established: 1964 (1980 reestablished as ICA-Working group)
Vocabularies:
Subject(s): cartography
UDC 528. 9 (084.3)
Language(s): Cs; Nl; En; Fr; De; Hu; It; Ja; Pl; Pt; Sk; Sv; 2nd edition: Ar; Ch; Da; Fi; No; Sr-Cr; Th
Publications: *Multilingual Dictionary of Technical Terms in Cartography/Dictionnaire multilingue des termes techniques cartographiques.* 1st edition. Franz Steiner Verlag, Wiesbaden 1973. LXXVIII, 573 pp.

(Project(s))
Title: Multilingual Dictionary of Technical Terms in Cartography, 2nd enlarged edition
Project type: commissioned project; in progress
Principal investigator: Prof. Dr. E. Meynen, Bonn; Prof. Dr. J. Neumann, Karlsruhe (Secretary)
Collaborators: Germany: 15; other national working groups: different number of persons
Scope and/or subject matter: Cartography, definitions in En, De, Es, Fr and Ru
Method of work: Teamwork of international and national working groups. Source language: De; other definition languages: En, Fr, Es and Ru ; equivalents in 9 other languages
Source material: Publications of official survey offices and other cartographical literature

○ ***100*** *(Organization)*
International Civil Aviation Organization/Organisation de l'aviation civile internationale
ICAO/OACI
1000 Sherbrooke St. W., Suite 400
P. O. Box 400 Phone: 285–8123–24–25
Montréal, Québec, H3A 2R2
Canada
Head: Dr. Assad Kotaite (Secretary General)
Expert(s): Mr. A. I. Oviedo
Staff categories:
35 (22 full time; 13 part time) 22 linguist(s)/terminologist(s) (special field); translators/interpreters co-operate with the Terminology Service in updating the ICAO-Lexicon in addition to consultants from universities, institutes, academies, laboratories and other terminological centres
Founded: 1947
Documentation: bibliographic data
Subject(s): aviation; air navigation; air transport; law
UDC 656. 7; UDC 656. 07; UDC 347. 8; UDC 629. 73
Language(s): En; Fr; Ru; Es; Ar
Vocabularies:
Subject(s): aviation; air navigation; air transport; law
UDC 656. 7; UDC 656. 07; UDC 347. 8; UDC 629. 73
Language(s): En; Fr; Ru; Es; Ar
Publications: *Terminology Bulletins* (glossaries of terms related to specific fields), primarily in En, Fr, Ru, Es; *Terminology Circulars*, in any ICAO languages; *Terminology Advisories*(guidelines for terminology standardization); (Please consult ICAO for a complete listing of publications)
Services: Various services are provided to other terminology bodies on an exchange basis

● ***101*** *(Commission/Committee)*
International Commission for Palynology. Working Group A 4 (Actuopalynology) on "Terminology of Pollen Grains and Spores"
ICP/A 4
Parent organization: International Commission for Palynology
435, Deike Building, Pennsylvania State University
Pennsylvania, PA, 16802
USA
Contact: Dr. Stephen Blackmore
Department of Botany
British Museum (Natural History)
Cromwell Road
London SW7 5BD
England
Chairperson: Dr. Stephen Blackmore
Established: 1973
Vocabularies:
Subject(s): pollen grains and spores
UDC 581. 46
Language(s): En; Fr
Vocabulary standards:
Subject(s): pollen grains and spores
UDC 581. 46
Language(s): En; Fr
Publications: *Nilsson, S.; Muller, J. Recommended Palynological Terms and Definitions,* 1978. (Grana 17: p.55–58) SS-58

● ***102*** *(Commission/Committee)*
International Commission on Stratigraphy
ICS (used only as part of document designations)
Parent organization: International Union of Geological Sciences
Dobelnsgatan 20 A
564 Phone: -46–18–13 44 48
Uppsala, S-751 22 Telex: 7604 UNIVUPS S
Sweden

Chairperson: Prof. Anders Martinsson, Chairman
Established: 1878 (reorganized 1952)
Vocabularies:
Subject(s): stratigraphy
UDC 551. 7
Language(s): En (and other languages)
Vocabulary standards: Subcommission on Stratigraphic Classification (Chairman: Prof.Amos Salvador)
Subject(s): stratigraphy
UDC 551. 7
Language(s): En (and other languages)
Theory: Subcommission on Stratigraphic Classification (Chairmann: Prof. Amos Salvador); En (and other languages)
Publications: Main document: *International Stratigraphic Guide*. A guide to stratigraphic classification, terminology and procedure. Wiley's 1976. Corresponding national documents exist as well

● *103* *(Commission/Committee)*
International Commission on Zoological Nomenclature/Commission internationale de nomenclature zoologique
ICZN; CINZ
Parent organization: International Union of Biological Sciences
51 Boulevard de Montmorency Phone: 525–00–09
Paris, F-75016
France
Contact: R. V. Melville, Esq., M. Sc.
International Commission on Zoological Nomenclature
c/o British Museum (Natural History)
Cromwell Road
London
SW7 5BD
England
Committee members: 26 (not ever fewer than 18)
Chairperson: Dr. C. W. Sabrosky
Vice-chairperson: Prof. Per Brinck
Established: 1895
Vocabularies:
Subject(s): zoology; palaeontology
UDC 59
Language(s): En; Fr
Vocabulary standards:
Subject(s): zoology; palaentology
UDC 59
Language(s): En; Fr; (Code is also translated into many different languages independent of ICZN).
Publications: *The Bulletin of Zoological Nomenclature; International Code of Zoological Nomenclature; Official Lists of Family, Genus and Specific Names in Zoology; Official Indexes of Family, Genus and Specific Names in Zoology*

● *104* *(Commission/Committee)*
International Committee for Unification of Terminological Neologisms - World Bank of International Terms
ICUTN/WBIT
Parent organization: International Federation of Translators/Fédération internationale des traducteurs
Al. Niepodległości 67 m. 69 Phone: 445561
Warszawa, 02 626
Poland
Committee members: 5 (members of ICUTN: appr.80)
Chairperson: Mr. Zygmunt Stoberski, M. A. FIL.
Vice-chairperson: Mr. Kazimierz Gebarski, M. A. (expert)
Established: 1978
Vocabularies:
Subject(s): neologisms (all fields)
UDC 801. 316. 1
Language(s): Ar; En; Fr; Ch; Hi; Sa; Ko; Pj; Be; Bg; Cs; Da; Et; Fi; De; Gr; Hu; It; La; Li; Nl; No; Pl; Pt; Ro; Ru; Sh; Sk; Sn; Es; Sv; Tr; Tk; Ew
Theory: Guidelines published in *"Babel"* No. 1/1975, No. 2/1979, No. 2/ 1980; in *"Le Langage et l'Homme"*, Bruxelles No. 37/1979 in *"Jista"*, New Delhi, Vol. 5, 1976; *Izvestija Akademii Nauk SSSR, Serija Literatury i Jazyka* tom 33, No. 5, 1974 etc.; En; Fr; Pl; Sh; Ru; etc.
Publications: *Babel*, Official Organ of the FIT, published with the assistance of Unesco from 1977 until now; Photocopies of column *International Scientific Terminology - Neologisms in the Course of Unification* published in *Babel* from 1977 onwards.

● *105* *(Organization)*
International Federation for Information Processing
IFIP
3, Rue du Marché
Genève, 1204
Switzerland
Staff: 2 (1 full-time, 1 part-time at IFIP Sect., Geneva)
Head: Prof. P. A. Bobillier, president
Expert(s): Dr. R. I. Tanaka, Chairman of International Liaison Committee
Founded: 1960
Vocabularies:
Subject(s): information processing
UDC 681. 3
Language(s): En; Fr; De; Nl; Sv; It; Ru; and others
Publications: Gould, I. H. (ed). *IFIP Guide to concepts and terms in data processing*. Amsterdam: North-Holland, 1971, 168 p.

(Project(s))
Title: IFIP Guide to concepts and terms in data processing
Project type: institutional; in planning
Scope and/or subject matter: Revision and updating of the 1971 edition of the IFIP Guide

● *106* *(Organization)*
International Information Centre for Terminology/ Centre international d'information pour la terminologie/Internationales Informationszentrum für Terminologie/Meždunarodnyj informacionnyj centr po terminologii
Infoterm
Parent organization: Österreichisches Normungsinstitut (ON) [Austrian Standards Institute]
Heinestraße 38
P. O. Box 130 Phone: (0222) 26 75 35
Wien, A-1021 Telex: 115960 onorm a
Austria Telegr.: Austrianorm
Staff: 7
Head: Prof. Helmut Felber, Director
Expert(s): Mr. Christian Galinski; Ms. Magdalena Krommer-Benz, M. A.; Mr. Adrian Manu; Mr. Wolfgang Nedobity, DipLib.
Staff categories:
1 terminologist(s) (theory and spezial field)/terminology teacher(s); 2 documentalist(s)/librarian(s) /linguist(s) (1 also managing function); 1 documentalist (s)/librarian(s)/translator(s); 1 terminologist(s) (special field)/documentalist(s)/librarian(s)/linguist(s)
Founded: 1971
Documentation:
bibliographic data
factographic data
Subject(s): all fields
UDC 0/9
Language(s): En; Fr; De; It; Es; Pt; Ru; Cs; Sh; Ro; Ja; Ch (and many others)
Vocabularies:
Subject(s): terminology; terminological data banks
UDC 001. 4 : 681. 3
Language(s): En; Fr; De
Vocabulary standards:
Subject(s): principles and methods of standardization
UDC 001. 4. 001
Language(s): En; Fr; De
Theory: En; De
Training: undergraduate; graduate; postgraduate
Access to training: available to outside users
Curriculum: fundamentals of terminology ; terminological principles; lexicographical methods

Publications: Wüster, E. *The Road to Infoterm.* Two reports prepared on behalf of Unesco. Inventory of Sources of Scientific and Technical Terminology. A Plan for Establishing an International Information Centre (Clearinghouse) for Terminology. München: Verlag Dokumentation, 1974, 141 p., A5. (Infoterm Series 1);
Felber, H.; Krommer-Benz, M.; Manu, A. *International Bibliography of Standardized Vocabularies/Bibliographie internationale de vocabulaires normalisés/Internationale Bibliographie der Normwörterbücher* (2nd edition). München/New York/Paris/London: K. G. Saur, 1979, 540 p., A4. (Infoterm Series 2); Supplement in print.
INFOTERM. *First Infoterm Symposium. International Co-operation in Terminology/Premier Symposium d'Infoterm. Coopération internationale en terminologie.* München: Verlag Dokumentation, 1976, 332 p., A5. (Infoterm Series 3);
Krommer-Benz, M. *World Guide to Terminological Activities/Guide mondial des activités terminologiques.* München: Verlag Dokumentation, 1977, 311 p., (Infoterm Series 4);
INFOTERM. *Terminological Data Banks.* Proceedings of the First International Conference convened in Vienna, April 2–3, 1979. München: Verlag Dokumentation, 1980, 207 p., A5. (Infoterm Series 5);
INFOTERM. *Theoretical and Methodological Problems of Terminology.* Proceedings of the International Symposium in Moscow, November 27–30, 1979. München: Verlag Dokumentation, 1981, 608 p., A5. (Infoterm Series 6);
INFOTERM. *Terminologies for the Eighties.* With a special section: 10 years of Infoterm. München/New York/London/Paris: K. G. Saur, 1982, 412 p., A5 (Infoterm Series 7), ISBN 3-598-21367-0, DM 86, -/US $ 48,-.
Infoterm Newsletter (published quarterly; available: nos. 1–33); *TermNet News* (published irregularly; available: No. 1, 2/3, 4/5, 6)
Felber, H.; Lang, F.; Wersig, G. (eds.) *Terminologie als angewandte Sprachwissenschaft.* Gedenkschrift für Univ.-Prof. Dr. Eugen Wüster [Terminology as applied linguistics. In memoriam Prof. Eugen Wüster]. München/New York/Paris/London: K. G. Saur, 1979, 272 p., A5;
Wüster, E. *Einführung in die allgemeine Terminologielehre und terminologische Lexikographie* [Introduction to the General Theory of Terminology and Terminological Lexicography]. Wien: Technische Universität, 1979, 2 vols., 215 p., A5. (Schriftenreihe der Technischen Universität Wien Band 8);
Fachsprache (co-editor); *Textes choisis de terminologie* (co-editor); articles in books and periodicals (list of documents available on request); International bibliographies of terminology (BTs): BT 1 (terminology); BT 2 (standardized vocabularies); BT 3 (multilingual specialized vocabularies); BT 4 (monolingual specialized vocabularies); BT 5 (World guide to terminological activities); BT 6 (guidelines); BT 7 (periodicals); BT 8 (bibliographies and catalogues); BT 9 (term banks); BT 10 (meetings); BT 11 (reports and proceedings); BT 12 (theses and dissertations); BT 13 (directories of acronyms, initialisms and abbreviations); BT 14 (collections of neologisms); BT 15 (Who's Who in Terminology)

Services: reference; in-house training; advising on the application of terminological methods and principles
No service charges

User aids: Infoterm Newsletter; TermNet News

Operation:
manual
computer-assisted (preparation of BTs (processing by Siemens, Munich; Université Laval, Canada; DGTD, Secretary of State, Canada; Institut für maschinelle Dokumentation (IMD) Graz, Austria))

Access:
via telephone/via telex

Data available:
on exchange basis (partly); with charge (partly); free of charge (partly)

(Project(s))
Title: Comparative study of terminological data elements
Project type: institutional; in progress
Principal investigator: Prof. H. Felber
Collaborators: 1
Scope and/or subject matter: Theory of terminology
Method of work: Theoretical approach
Source material: ISO/TC 37 documents; E. Wüster's "Theory of terminology"; documents issued by terminological data banks; TermNet 4(9)-79; TermNet 4–80
Publications: Felber, H. *A study of terminological data elements on the basis of the general theory of terminology and a comparative study of those data elements used in terminological data banks all over the world.* Vienna: Infoterm, 1980, 9p. + 12p. + 8p., A4. (Draft. TermNet 1–80) Felber, H. *Guidelines for the recording of terminological data for machine processing.* Vienna: Infoterm, 1982, 28p., A4. (Draft.TermNet 7–82).

Title: Computerunterstütztes Fachwortsuchsystem für Japanisch [Computer Aided Retrieval System for Specialized Japanese Vocabularies]
Project type: institutional
Principal investigator: Dr. E. Simoncsics; Ch. Galinski
Collaborators: 8
Scope and/or subject matter: Terminology of architecture
Method of work: Coding of Chinese ideographs in the Japanese writing system according to technical criteria without neglecting the etymology. Japanese vocabulary is microfilmed and retrieved in a computerized microfiche retrieval system by use of this code.
Source material: Japanese specialized vocabularies concerning architecture
Publications: Working papers

Title: Directory of Terminological Meetings (BT 10)
Project type: institutional; in progress
Principal investigator: W. Nedobity
Collaborators: 2
Scope and/or subject matter: A list of all past and forthcoming meetings with topics relevant to terminology
Method of work: Collection of data on meetings
Source material: Journals, invitations and other announcements as well as proceedings, reports, etc.
Publications: *Infoterm Newsletter* 1–22 ff, especially INL 9, published in Lebende Sprachen 23 (1978), no. 1, p. 42–43

Title: International Bibliography of Bibliographies and Catalogues Containing Terminological Literature (BT 8)
Project type: institutional; in progress
Principal investigator: A. Manu
Collaborators: 2
Scope and/or subject matter: Bibliographies of terminological literature, of dictionaries, vocabularies and thesauri
Method of work: Collection and analysis of bibliographies
Source material: Catalogues of publishing houses, etc.
Publications: *Infoterm Newsletter* 2 published in *Lebende Sprachen 21* (1976), no. 4, p. 189; *Infoterm Newsletter* 22 publ. in *International Classification* 9 (1982) no. 1, p. 31–32.

Title: International Bibliography of Collections of Neologisms (BT 14)
Project type: institutional; in progress
Collaborators: 2
Scope and/or subject matter: Recording of listings of new concepts and terms pertaining to all subjects and languages
Method of work: Collection and analysis of subject vocabularies specializing in new terms
Source material: Scientific journals, brochures, books

Title: International Bibliography of Directories of Acronyms, Initialisms and Abbreviations (BT 13)
Project type: institutional; in progress
Collaborators: 2
Scope and/or subject matter: Recording of the bibliogra-

phic data of lists giving the full forms of abbreviations of terms and names
Method of work: Scientific journals, books, bibliographies

Title: International Bibliography of Monolingual Specialized Vocabularies (BT 4)
Project type: institutional; in planning
Principal investigator: A. Manu
Collaborators: 3
Scope and/or subject matter: It pertains to all subject fields and to all languages
Method of work: Collecting and recording of monolingual subject vocabularies
Source material: Other bibliographies, library and commercial catalogues

Title: International Bibliography of Multilingual Specialized Vocabularies (BT 3)
Project type: institutional; in planning
Principal investigator: A. Manu
Collaborators: 3
Scope and/or subject matter: It pertains to all subject fields and to all languages
Method of work: Collecting and recording of multilingual subject vocabularies
Source material: Other bibliographies, library and commercial catalogues

Title: International Bibliography of Periodicals Pertaining to Terminology (BT 7)
Project type: institutional; in progress
Collaborators: 2
Scope and/or subject matter: All periodical publications pertaining to terminology and related subjects
Method of work: Collection and analysis of terminological journals
Source material: Various bibliographies
Publications: *Infoterm Newsletter* 3 published in *Lebende Sprachen* 22 (1977), no. 1, p. 46; *Infoterm Newsletter* 5 published in Lebende Sprachen 22 (1977), no. 3, p. 144; *Infoterm Newsletter* 21 published in *International Classification* 8 (1981), no. 3, p. 156

Title: International Bibliography of Reports and Proceedings of Terminological Meetings (BT 11)
Project type: institutional; in progress
Principal investigator: W. Nedobity
Collaborators: 2
Scope and/or subject matter: All literature pertaining to terminological meetings or meetings on related subjects
Method of work: Collection and analysis of relevant literature
Source material: Announcements in journals and bibliographies

Title: International Bibliography of Standards and Nonstandardized Guidelines for Terminology (BT 6)
Project type: institutional; in progress
Collaborators: 2
Scope and/or subject matter: Terminology; terminography
Method of work: Collection and bibliographic control of standards
Source material: International and national standards on terminological principles; standard-type documents
Publications: *Infoterm Newsletter* 4. Published in *Lebende Sprachen* 22 (1977), no. 2, p. 93–95; *Infoterm Newsletter* 20. Published in *International Classification* 8 (1981), no. 2, p. 98–100; *Infoterm Newsletter* 4 (revised)

Title: International Bibliography of Terminological Literature (BT 1)
Project type: institutional; in progress
Principal investigator: Prof. G. Rondeau (GIRSTERM); Prof. H. Felber (Infoterm)
Collaborators: 2
Scope and/or subject matter: Terminology science; terminology work; terminography
Method of work: Collection and analysis of all bibliographical references of works pertaining to terminology
Source material: Reports; studies; papers; articles; monographs
Publications: RONDEAU, G.; FELBER, H. *Bibliographie internationale de la terminologie. Exposé de la méthode.* 2e éd. Québec: GIRSTERM/Université Laval, 1981, 68 p., 215x280mm, ISBN 2-920242-02-4.
International Bibliography of terminological literature. Quebec: Université Laval, forthcoming; Reseach report and computer print-out of GIRSTERM

Title: International Bibliography of Terminological Theses and Dissertations (BT 12)
Project type: institutional; in progress
Collaborators: 2
Scope and/or subject matter: All terminological works submitted for academic awards
Method of work: Collaboration with universities and other academic institutions
Publications: A special worksheet has been designed

Title: International Bibliography on Computer-aided Terminology (BT 9)
Project type: institutional; in progress
Principal investigator: M. Krommer-Benz
Collaborators: 2
Scope and/or subject matter: All literature dealing with the recording and machineprocessing of terminological data
Method of work: Collection and analysis of relevant literature
Source material: Publications of terminological data banks, universities, etc.
Publications: *Infoterm Newsletter 1*, published in *Lebende Sprachen* 21 (1976), no. 3, p. 140–142; revised publication pending.

Title: Internationaler Terminologieschlüssel [Key to international terminology]
Project type: institutional; in planning
Principal investigator: W. Nedobity
Collaborators: 1
Scope and/or subject matter: International unification of concepts and terms; formation of terms with international elements
Method of work: Theoretical and practical approach
Source material: E. Wüster's "Internationaler Terminologieschlüssel" [Key to international terminology] (unpublished)

Title: Model Terminology Data Bank
Project type: institutional; in progress
Collaborators: 4
Scope and/or subject matter: A pilot project for the input, processing and retrieval of terminological data

Title: Updating of the International Bibliography of Standardized Vocabularies (BT 2)
Project type: institutional; in progress
Principal investigator: A. Manu
Collaborators: 3
Scope and/or subject matter: List of all terminological standards of the world
Method of work: Collection and analysis of terminological standards
Source material: Catalogues of standards institutions; collection of foreign standards at the Austrian Standards Institute

Publications: FELBER, H.; KROMMER-BENZ, M.; MANU, A. *International Bibliography of Standardized Vocabularies/Bibliographie internationale de vocabulaires normalisés/Internationale Bibliographie der Normwörterbücher.* München/ New York/London/Paris: K. G. Saur, 2nd edition, 1979, 540 p., A4. (Infoterm Series 2) Supplement forthcoming; Computer print-out

Title: Updating of the World Guide to Terminological Activities (BT 5)
Project type: institutional; in progress
Principal investigator: M. Krommer-Benz, M. A.
Collaborators: 2
Scope and/or subject matter: Information on terminological activities pertaining to all subjects and languages
Method of work: Survey via questionnaires; analysis
Source material: Questionnaires; informants; documents; first edition of the "World Guide to Terminological Activities"
Publications: KROMMER-BENZ, M. *World Guide to Terminological Activities/Guide mondial des activités terminologiques.* München: Verlag Dokumentation, 1976, 332 p., A5. (Infoterm Series 4) New edition expected in 1984; Computer print-out
Title: Vocabulary of terminological data banks (VTDB)
Project type: institutional; in planning
Principal investigator: I. Hohnhold (Siemens); W. Nedobity (Infoterm)
Collaborators: 8
Method of work: Evaluation of all literature and reference material on terminological data banks, in English, French and German
Publications: Computer print-out of Siemens, Munich, *Vocabulary of terminological data banks*

Title: Vocabulary of Terminology - General
Project type: institutional; in progress
Principal investigator: W. Nedobity
Scope and/or subject matter: Systematic vocabulary of terminology science and work
Source material: Wüster, E. Wörterbuch des Benennungswesens [Vocabulary of terminology science]; ISO/R 1087 "Vocabulary of terminology"
Publications: Computer print-out of a *Vocabulary of terminological data banks*

Title: Who's Who in Terminology (BT 15)
Project type: institutional; in planning
Collaborators: 2
Scope and/or subject matter: Directory of terminology experts and executives engaged in terminology work
Method of work: Biographical inquiry
Source material: Questionnaires
Publications: *Who's Who in Terminology*

● ***107****(Organization)*

International Institute of Welding/Institut international de la soudure
IIW/IIS

54 Princes Gate,
Exhibition Road
London, SW7 2PG
UK
Phone: 01-584-8556
Telex: 81183
Telegr.: Soudweld, London SW7

Head: Mr. P. D. Boyd, Secretary General
Expert(s): Prof. Dr. P. Štular, Chairman of Commission VI "Terminology", Institut za varilstvo, 61000 Ljubljana, Ptujska 19, Yugoslavia
Founded: 1948
Documentation: factographic data
Subject(s): welding; brazing; soldering; braze welding; thermal cutting; hot spraying; ultrasonic testing
UDC 621. 791
Language(s): Cs; Da; Nl; En; Fi; Fr; De; It; Pl; Ro; Ru; Sr; Sk; Sn; Es; Sv; Tr; No
Vocabularies:
Subject(s): welding; brazing; soldering; braze welding; thermal cutting; hot spraying; ultrasonic testing
UDC 621. 791
Language(s): Cs; Da; Nl; En; Fi; Fr; De; It; No; Pl; Ro; Ru; Sr; Sk; Sn; Es; Sv; Tr
Vocabulary standards:
Subject(s): welding; brazing; soldering
UDC 621. 791
Language(s): En/Fr
Publications: Multilingual Collection of Terms:
Section I *General Terms - Welding procedure, characterisytics and inspection of welds*; Section II *Gas welding* (revised edition); Section III *Arc welding*; Section IV *Resistance welding*; Section V *Thermal cutting*; Section VI *Hot spraying*; Section VII *Brazing, soldering and braze welding*; Section VIII *Welding of plastics*; Section IX *Special welding processes*; List of terms used in ultrasonic testing; bi-monthly journal *Welding in the World/Le Soudage dans le Monde*; numerous other publications not relevant to terminology

(Project(s))
Title: Continuation of the revision and publication of the booklet "General Terms" of the Multilingual Collection of Terms
Project type: institutional; in progress
Principal investigator: Prof. Dr. P. Štular
Collaborators: 25
Scope and/or subject matter: Revision of the existing sections of the Multilingual Collection of Terms to include definitions
Method of work: Collaborative work of experts appointed by the different national delegations represented in Commission VI

Title: Coordination of the work concerning terminology with other IIW Commissions and other organisms interested in welding terminology
Project type: institutional; in progress
Principal investigator: Prof. Dr. P. Štular
Collaborators: 25
Scope and/or subject matter: Revision of the existing sections of the Multilingual Collection of Terms to include definitions
Method of work: Collaborative work of experts appointed by the different national delegations represented in Commission VI

Title: Publication of the booklet "Gas welding" of the Multilingual Collection of Terms in 1982
Project type: institutional; in progress
Principal investigator: Prof. Dr. P. Štular
Collaborators: 25
Scope and/or subject matter: Revision of the existing sections of the Multilingual Collection of Terms to include definitions
Method of work: Collaborative work of experts appointed by the different national delegations represented in Commission VI

Title: Publication of the booklet "Special welding processes" of the Multilingual Collection of Terms
Project type: institutional; in progress
Principal investigator: Prof. Dr. P. Štular
Collaborators: 25
Scope and/or subject matter: Revision of the existing sections of the Multilingual Collection of Terms to include definitions
Method of work: Collaborative work of experts appointed by the different national delegations represented in Commission VI

Title: Review of the "Classification of welding processes - Terms and Definitions" booklet

Project type: institutional; in progress
Principal investigator: Prof. Dr. P. Štular
Collaborators: 25
Scope and/or subject matter: Revision of existing sections of the Multilingual Collection of Terms to include definitions
Method of work: Collaborative work of experts appointed by the different national delegations represented in Commission VI

Title: Revision and publication of the booklet "Arc welding" of the Multilingual Collection of Terms
Project type: institutional; in progress
Principal investigator: Prof. Dr. P. Štular
Collaborators: 25
Scope and/or subject matter: Revision of the existing sections of the Multilingual Collection of Terms to include definitions
Method of work: Collaborative work of experts appointed by the different national delegations represented in Commission VI

Title: Revision and publication of the booklet "Resistance welding" of the Multilingual Collection of Terms
Project type: institutional; in progress
Principal investigator: Prof. Dr. P. Štular
Collaborators: 25
Scope and/or subject matter: Revision of the existing sections of the Multilingual Collection of Terms to include definitions
Method of work: Collaborative work of experts appointed by the different national delegations represented in Commission VI

Title: Settlement of definitions for main welding, brazing and soldering processes
Project type: institutional; in progress
Principal investigator: Prof. Dr. P. Štular
Collaborators: 25
Scope and/or subject matter: Revision of the existing sections of the Mulilingual Collections of Terms to include definitions
Method of work: Collaborative work of experts appointed by the different national delegations represented in Commission VI

Title: Settlement of the booklet "Adhesive bonding of metals and plastics" of the Multilingual Collection of Terms
Project type: institutional; in progress
Principal investigator: Prof. Dr. P. Štular
Collaborators: 25
Scope and/or subject matter: Revision of the existing sections of the Multilingual Collection of Terms to include definitions
Method of work: Collaborative work of experts appointed by the different national delegations represented in Commission VI

● ***108*** *(Commission/Committee)*
Commission VI "Terminology/Terminologie"
Parent organization: International Institute of Welding/Institut International de la Soudure
54 Princes Gate,
Exhibition Road
London, SW7 2PG
UK
Phone: 01–584–8556
Telex: 81183
Telegr.: Soudweld, London SW7
Contact: Prof. Dr. P. Štular
Institut za varilstvo
61000 Ljubljana
Ptujska 19, Yugoslavia
Committee members: 25
Chairperson: Prof. Dr. P. Štular
Vice-chairperson: Mr. C. Poisson
Established: 1949
Vocabularies:
Subject(s): welding; brazing; soldering; braze welding; thermal cutting; hot spraying; ultrasonic testing
UDC 621. 791
Language(s): Cs; Da; Nl; En; Fi; Fr; De; It; No; Pl; Ro; Ru; Sr; Sk; Sn; Es; Sv; Tr
Vocabulary standards:
Subject(s): welding; brazing; soldering;
UDC 621. 791
Language(s): En/Fr

● ***109*** *(Organization)*
International Monetary Fund
[Fonds monétaire international; Fondo Monetario Internacional (Es)]
IMF; FMI; FMI (Es)
700, 19th Street NW
Washington, D. C., 20431
USA
Phone: (202) 477–7000
Telex: ITT 440040 Fund UI
Telegr.: Interfund - Washington DC
Head: Jacques de Larosière, Managing Director
Expert(s): Georges G. Van Massenhove
Founded: 1945
Documentation: bibliographic data
Subject(s): economics; finance
UDC 33
Language(s): En; Fr; Es
Vocabularies:
Subject(s): economics; finance; fiscal matters
UDC 33
Language(s): En; Fr; Es

● ***110*** *(Organization)*
International Organization for Human Ecology/Internationale Organisation für Humanökologie/Organisation internationale pour l'écologie humaine/Organizzazione internazionale per l'ecologia umana/Organización internacional para la ecología humana
I. O. H. E.
Karlsplatz 13
Wien, A-1040
Austria
Phone: 65 37 85/ext. 799
Staff: 6 (working with Subgroup "Terminology" on a voluntary basis)
Head: Univ. Prof. Dr. Robert Reichardt (President)
Expert(s): Dr. Helmut Knötig
Staff categories:
1 terminologist(s) (theory and special field); 5 other professional(s) (natural and social scientists)
Founded: 1960 (1970)
Documentation: factographic data
Subject(s): human ecology
UDC 574
Language(s): De; En; Fr
Vocabularies:
Subject(s): human ecology
UDC 574
Language(s): De; En; Fr
Publications: Bimonthly (since 1976) *Colloquium internationale*; in De, En, Fr; Ed.: H. Knötig, Vienna; Publishers: Archivum Oecologiae hominis;
Quarterly *Folia Oecologiae hominis/Humanökologische Blätter*; in De, En; Ed.: H. Knötig, Vienna; Publishers: Archivum Oecologiae hominis; Knötig, H. (ed.) Proceedings of Vienna International Meetings on Human Ecology. Vienna: Archivum Oecologiae hominis.

(Project(s))
Responsible: Subgroup "Terminology"
Title: Updating of the terminology of human ecology
Project type: commissioned project: (International Organization for Human Ecology); in progress
Principal investigator: Dr. Helmut Knötig
Collaborators: 6
Scope and/or subject matter: Terminology of human ecology

Method of work: Compilation of terms on the basis of existing and available material
Source material: Material available
Publications: *Colloquium internationale*; in De, En, Fr; Ed.: H. Knötig; Vienna; Publishers: Archivum Oecologiae hominis

● *111* *(Commission/Committee)*
Subgroup "Terminology"
Parent organization: International Organization for Human Ecology
Karlsplatz 13 Phone: 65 37 85/ext. 799
Wien, A-1040
Austria
Committee members: 6
Chairperson: Dr. H. Knötig
Established: Predecessor "Commission on Terminology", founded 1975; Working Group "Terminology": 1977; Subgroup "Terminology": 1979
Vocabularies:
Subject(s): human ecology
UDC 574
Language(s): De; En; Fr
Publications: Bimonthly (since 1976) *Colloquium internationale*; in De, En, Fr; Ed.: H. Knötig; Vienna; Publishers: Archivum Oecologiae hominis; Quarterly *Folia Oecologiae hominis/Humanökologische Blätter*; in De, En; Ed,: H. Knötig; Publishers: Archivum Oecologiae hominis; Knötig, H. (ed.) Proceedings of Vienna International Meetings on Human Ecology. Vienna: Archivum oecologiae hominis.

● *112* *(Organization)*
International Organization for Standardization/ Organisation internationale de normalisation/ Meždunarodnaja Organizacija po Standartizacii
ISO
1, rue de Varembé
P. O. Box 56 Phone: (022) 34 12 40
Genève 20, CH-1211 Telex: 23 88 7 ISO CH
Switzerland Telegr.: Isorganiz
Head: Mr. D. C. Kothari, President; M. Olle Sturen, Secretary General
Expert(s): (Actual terminology work is being carried out within Technical Committees by Sub-Committees and Working Groups)
Founded: 1947/Successor of the International Federation of National Standardizing Associations (ISA)
Documentation:
bibliographic data
factographic data
Subject(s): all fields (except electrical and electronic engineering)
UDC 0/9
Language(s): En; Fr; Ru; additional languages in exceptional cases
Vocabulary standards:
Subject(s): all fields (except electrical and electronic engineering)
UDC 0/9
Language(s): En; Fr; Ru
Theory: Within ISO/TC 37; En; Fr; Ru
Publications: International Standards; ISO Catalogue; ISO Technical programme; ISO Standards Handbooks;
ISO Bibliographies: *Bibliography 7 - Documentation and terminology; Bibliography 8 - International Standard Vocabularies*;
ISO Annual Review; ISO Bulletin; ISO Memento; ISO Member Bodies; ISO Liaisons; Directory of international standardizing bodies; ISO/IEC technical collaboration; ISO Participation; ISO Constitution and rules of procedure; ISO Council resolutions; Directives for the technical work of ISO; ISONET Constitution; The aims and principles of standardization; Pamphlets and leaflets; Listings (twice a year)
Services: reference (without charge); referral (without charge); other: sale of standards
User aids: ISO Bulletin, published monthly; ISO Catalogue, published annually; ISO bibliographies

● *113* *(Commission/Committee)*
TC 4 "Rolling bearings", WG 8 "Vocabulary"/TC 4 "Roulements", GT 8 "Vocabulaire"
ISO/TC 4/WG 8
Parent organization: International Organization for Standardization
Contact: Association française de normalisation (AFNOR)
Tour Europe
Cedex 7
92080 Paris La Défense, France
Vocabulary standards:
Subject(s): rolling bearings
UDC 621. 822. 6/. 8
Language(s): En; Fr; Ru

● *114* *(Commission/Committee)*
TC 6 "Paper, board and pulps", SC 1 "Nomenclature, terminology, substances"/TC 6 "Papiers, cartons et pâtes", SC 1 "Nomenclature, terminologie, masses"
ISO/TC 6/SC 1
Parent organization: International Organization for Standardization
Contact: Association française de normalisation (AFNOR)
Tour Europe
Cedex 7
92080 Paris La Défense, France
Vocabulary standards:
Subject(s): paper; board and pulps
UDC 676
Language(s): En; Fr; Ru

● *115* *(Commission/Committee)*
TC 6 "Paper, board and pulps", SC 1 "Nomenclature, terminology, substances", WG 1 "Revision and coordination of ISO terminology for paper"/TC 6 "Papiers, cartons et pâtes", SC 1 "Nomenclature, terminologie, masses", GT 1 "Révision et coordination de la terminologie ISO du papier"
ISO/TC 6/SC 1/WG 1
Parent organization: International Organization for Standardization
Contact: British Standards Institution (BSI)
2 Park Street
London W1A 2BS
UK
Vocabulary standards:
Subject(s): paper; board and pulps
UDC 676
Language(s): En; Fr; Ru

● *116* *(Commission/Committee)*
TC 6 "Paper, board and pulps", SC 6 "Envelopes and postcards", WG 1 "Envelopes, pockets and postcards - Terminology"/TC 6 "Papiers, cartons et pâtes", SC 6 "Enveloppes et cartes postales", GT 1 "Enveloppes, pochettes, cartes postales - Terminologie"
ISO/TC 6/SC 6/WG 1
Parent organization: International Organization for Standardization
Contact: British Standards Institution (BSI)
2 Park Street
London W1A 2BS
UK
Vocabulary standards:
Subject(s): paper; board and pulps
UDC 676
Language(s): En; Fr; Ru

● *117* *(Commission/Committee)*
TC 8 "Shipbuilding and marine structures", SC 8 "Ships side scuttles and rectangular windows", WG 7 "Terminology"/TC 8 "Construction navale et structures maritimes", SC 8 "Hublots et fenêtres rectangulaires", GT 7 "Terminologie"
ISO/TC 8/SC 8/WG 7
Parent organization: International Organization for Standardization
Contact: British Standards Institution (BSI)
2 Park Street
London W1A 2BS
U. K.
Vocabulary standards:
Subject(s): shipbuilding; marine structures
UDC 629. 12
Language(s): En; Fr; Ru

● *118* *(Commission/Committee)*
TC 8 "Shipbuilding and marine structures", SC 11 "Terminology, symbols, drawings, etc." /TC 8 "Construction navale et structures maritimes", SC 11 "Terminologie, symboles, dessins, etc."
ISO/TC 8/SC 11
Parent organization: International Organization for Standardization
Contact: India Standards Institution (ISI)
Manak Bhavan
9 Bahadur Shah Zafar Marg
New Delhi 110002
India
Vocabulary standards:
Subject(s): shipbuilding; marine structures
UDC 629. 12
Language(s): En; Fr; Ru

● *119* *(Commission/Committee)*
TC 8 "Shipbuilding and marine structures", SC 15 "Computer applications in shipbuilding", WG 1 "Terminology and definitions"/TC 8 "Construction navale et structures maritimes", SC 15 "Applications des ordinateurs en construction navale", GT 1 "Terminologie et définitions"
ISO/TC 8/SC 15/WG 1
Parent organization: International Organization for Standardization
Contact: British Standards Institution (BSI)
2 Park Street
London W1A 2BS
UK
Vocabulary standards:
Subject(s): shipbuilding; marine structures
UDC 629. 12
Language(s): En; Fr; Ru

● *120* *(Commission/Committee)*
TC 10 "Technical drawings", SC 5 "Dimensioning and tolerancing", WG 7 "Terminology"/TC 10 "Dessins techniques", SC 5 "Cotation et tolérancement", GT 7 "Terminologie"
ISO/TC 10/SC 5/WG 7
Parent organization: International Organization for Standardization
Contact: Vacant
Vocabulary standards:
Subject(s): technical drawings
UDC 744. 4
Language(s): En; Fr; Ru

● *121* *(Commission/Committee)*
TC 10 "Technical drawings", SC 8 "Building drawings", WG 10 "Terminology"/TC 10 "Dessins techniques", SC 8 "Dessins de bâtiment", GT 10 "Terminologie"
ISO/TC 10/SC 8/WG 10
Parent organization: International Organization for Standardization
Contact: Vacant
Vocabulary standards:
Subject(s): technical drawings
UDC 744. 4
Language(s): En; Fr; Ru

● *122* *(Commission/Committee)*
TC 17 "Steel", SC 2 "Terminology, classification and designation of steel"/TC 17 "Acier", SC 2 "Terminologie, classification et désignation des aciers"
ISO/TC 17/SC 2
Parent organization: International Organization for Standardization
Contact: India Standards Institution (ISI)
Manak Bhavan
9 Bahadur Shah Zafar Marg
New Delhi 110002
India
Vocabulary standards:
Subject(s): steel
UDC 669. 14
Language(s): En; Fr; Ru

● *123* *(Commission/Committee)*
TC 20 "Aircraft and space vehicles", SC 3 "Terms and symbols for flight dynamics"/TC 20 "Aéronautique et espace", SC 3 "Termes et symboles de la mécanique du vol"
ISO/TC 20/SC 3
Parent organization: International Organization for Standardization
Contact: Association française de normalisation (AFNOR)
Tour Europe
Cedex 7
92080 Paris La Défense
France
Vocabulary standards:
Subject(s): aircraft; space vehicles
UDC 629. 7
Language(s): En; Fr; Ru

● *124* *(Commission/Committee)*
TC 20 "Aircraft and space vehicles", SC 8 "Aerospace terminology"/TC 20 "Aéronautique et espace", SC 8 "Terminologie aéronautique"
ISO/TC 20/SC 8
Parent organization: International Organization for Standardization
Contact: USSR State Committee for Standards (GOST)
Leninsky Prospekt 9
Moskva 117049
USSR
Vocabulary standards:
Subject(s): aircraft; space vehicles
UDC 629. 7
Language(s): En; Fr; Ru

● *125* *(Commission/Committee)*
TC 20 "Aircraft and space vehicles", SC 10 "Aerospace fluid systems and components", WG 10 "Terminology"/TC 20 "Aéronautique et espace", SC 10 "Systèmes aérospatiaux de fluides et éléments constitutifs", GT 10 "Terminologie"
ISO/TC 20/SC 10/WG 10
Parent organization: International Organization for Standardization
Contact: DIN Deutsches Institut für Normung
Burggrafenstraße 4–10
Postfach 1107
D-1000 Berlin 30, FRG

Vocabulary standards:
Subject(s): aircraft; space vehicles
UDC 629. 7
Language(s): En; Fr; Ru

● ***126***(*Commission/Committee*)
TC 21 "Equipment for fire protection and fire fighting", SC 1 "Terminology, symbols, signs and classification of fire"/TC 21 "Equipment de protection et de lutte contre l'incendie", SC 1 "Terminologie, symboles, signes et classes de feux"
ISO/TC 21/SC 1
Parent organization: International Organization for Standardization
Contact: Association française de normalisation (AFNOR)
Tour Europe
Cedex 7
92080 Paris La Défense
France
Vocabulary standards:
Subject(s): fire protection; fire fighting
UDC 614. 84
Language(s): En; Fr; Ru

● ***127***(*Commission/Committee*)
TC 22 "Road vehicles", SC 1 "Ignition equipment", WG 1 "Terminology"/TC 22 "Véhicules routiers", SC "Equipement d'allumage", GT 1 "Terminologie"
ISO/TC 22/SC 1/WG 1
Parent organization: International Organization for Standardization
Contact: American National Standards Institute (ANSI)
1430 Broadway
New York
N. Y. 10018
USA
Vocabulary standards:
Subject(s): road vehicles
UDC 629. 11
Language(s): En; Fr; Ru

● ***128***(*Commission/Committee*)
TC 22 "Road vehicles", SC 6 "Terms and definitions of dimensions and weights"/TC 22 "Véhicules routiers", SC 6 "Dénominations et définitions des dimensions et des poids"
ISO/TC 22/SC 6
Parent organization: International Organization for Standardization
Contact: Ente Nazionale Italiano di Unificazione (UNI)
Piazza Armando Diaz 2
20123 Milano
Italy
Vocabulary standards:
Subject(s): road vehicles
UDC 629. 11
Language(s): En; Fr; Ru

● ***129***(*Commission/Committee*)
TC 22 "Road vehicles", SC 7 "Injection equipments and filters for use on road vehicles", WG 4 "Nomenclature and terminology"/TC 22 "Véhicules routiers", SC 7 "Equipements d'injection et filtres pour application aux véhicules routiers", GT 4 "Nomenclature et terminologie"
ISO/TC 22/SC 7/WG 4
Parent organization: International Organization for Standardization
Contact: British Standards Institution (BSI)
2 Park Street
London W1A 2BS
UK
Vocabulary standards:
Subject(s): road vehicles
UDC 629. 11
Language(s): En; Fr; Ru

● ***130***(*Commission/Committee*)
TC 22 "Road vehicles", SC 8 "Lighting and signalling", WG 5 "Terminology for lighting and light signalling devices on road vehicles"/TC 22 "Véhicules routiers", SC 8 "Eclairage et signalisation", GT 5 "Terminologie intéressant l'éclairage et la signalisation lumineuse des véhicules routiers"
ISO/TC 22/SC 8/WG 5
Parent organization: International Organization for Standardization
Contact: Ente Nazionale Italiano di Unificazione (UNI)
Piazza Armando Diaz 2
20123 Milano
Italy
Vocabulary standards:
Subject(s): road vehicles
UDC 629. 11
Language(s): En; Fr; Ru

● ***131***(*Commission/Committee*)
TC 22 "Road vehicles", SC 9 "Vehicle dynamics and road holding ability", WG 1 "Terminology"/TC 22 "Véhicules routiers", SC 9 "Dynamique des véhicules et tenue de route", GT 1 "Terminologie"
ISO/TC 22/SC 9/GT 1
Parent organization: International Organization for Standardization
Contact: Association française de normalisation (AFNOR)
Tour Europe
Cedex 7
92080 Paris La Défense
France
Vocabulary standards:
Subject(s): road vehicles
UDC 629. 11
Language(s): En; Fr; Ru

● ***132***(*Commission/Committee*)
TC 22 "Road vehicles", SC 13 "Ergonomics applicable to road vehicles", WG 4 "Terminology and definitions"/ TC 22 "Véhicules routiers", SC 13 "Ergonomie applicable aux véhicules routiers", GT 4 "Terminologie et définitions"
ISO/TC 22/SC 13/WG 4
Parent organization: International Organization for Standardization
Contact: Association française de normalisation (AFNOR)
Tour Europe
Cedex 7
92080 Paris La Défense
France
Vocabulary standards:
Subject(s): road vehicles
UDC 629. 11
Language(s): En; Fr; Ru

● ***133***(*Commission/Committee*)
TC 22 "Road vehicles", SC 21 "Electric road vehicles", WG 2 "Terminology. Definitions and methods of measurement of vehicle performance and of energy consumption"/TC 22 "Véhicules routiers", SC 21 "Véhicules électriques routiers", GT 2 "Terminologie. Définitions et méthodes de mesure des performances et de la consommation d'énergie"
ISO/TC 22/SC 21/WG 2
Parent organization: International Organization for Standardization
Contact: Association française de normalisation (AFNOR)
Tour Europe

Cedex 7
92080 Paris La Défense
France
Vocabulary standards:
Subject(s): road vehicles
UDC 629. 11
Language(s): En; Fr; Ru

● *134* *(Commission/Committee)*
TC 22 "Road vehicles", SC 23 "Mopeds", WG 4 "Terminology"/TC 22 "Véhicules routiers", SC 23 "Cyclomoteurs", GT 4 "Terminologie"
ISO/TC 22/SC 23/WG 4
Parent organization: International Organization for Standardization
Contact: Association française de normalisation (AFNOR)
Tour Europe
Cedex 7
92080 Paris la Défense
France
Vocabulary standards:
Subject(s): road vehicles
UDC 629. 11
Language(s): En; Fr; Ru

● *135* *(Commission/Committee)*
TC 23 "Tractors and machinery for agriculture and forestry", SC 1 "Terminology"/TC 23 "Tracteurs et matériels agricoles et forestiers", SC 1 "Terminologie"
ISO/TC 23/SC 1
Parent organization: International Organization for Standardization
Contact: Association française de normalisation (AFNOR)
Tour Europe
Cedex 7
92080 Paris La Défense
France
Vocabulary standards:
Subject(s): tractors; machinery for agriculture and forestry
UDC 631. 3; UDC 634. 0. 36/. 37
Language(s): En; Fr; Ru

● *136* *(Commission/Committee)*
TC 23 "Tractors and machinery for agriculture and forestry", SC 5 "Equipment for working the soil", WG 2 "Terminology for soil working elements"/TC 23 "Tracteurs et matériels agricoles et forestiers", SC 5 "Matériel de travail du sol", GT 2 "Terminologie des éléments pour le travail du sol"
ISO/TC 23/SC 5/WG 2
Parent organization: International Organization for Standardization
Contact: USSR State Committee for Standards (GOST)
Leninsky Prospekt 9
Moskva 117049
USSR
Vocabulary standards:
Subject(s): tractors; machinery for agriculture and forestry
UDC 631. 3; UDC 634. 0. 36/. 37
Language(s): En; Fr; Ru

● *137* *(Commission/Committee)*
TC 23 "Tractors and machinery for agriculture and forestry", Sc 1 "Terminology", WG 1 "Classification"/TC 23 "Tracteurs et matériels agricoles et forestiers", SC 1 "Terminologie", GT 1 "Classification"
ISO/TC 23/SC 1/WG 1
Parent organization: International Organization for Standardization
Contact: Instituto Nacional de Racionalización y Normalización (IRANOR)
Zurbano 46
Madrid 10
Spain
Vocabulary standards:
Subject(s): tractors; machinery for agriculture and forestry
UDC 631. 3; UDC 634. 0. 36/. 37
Language(s): En; Fr; Ru

● *138* *(Commission/Committee)*
TC 24 "Sieves, sieving and other sizing methods", SC 6 "Terminology"/TC 24 "Tamis, tamisage et autres méthodes de séparation granulométrique", SC 6 "Terminologie"
ISO/TC 24/SC 6
Parent organization: International Organization for Standardization
Contact: Standards Council of Canada (SCC)
International Standardization Branch
2000 Argentia Road
Suite 2-401
Mississauga
Ontario
L5N 1V8
Canada
Vocabulary standards:
Subject(s): sieves; sizing methods
UDC 621. 928
Language(s): En; Fr; Ru

● *139* *(Commission/Committee)*
TC 26 "Copper and copper alloys", WG 5 "Terminology"/TC 26 "Cuivre et alliages de cuivre", GT 5 "Terminologie"
ISO/TC 26/WG 5
Parent organization: International Organization for Standardization
Contact: DIN Deutsches Institut für Normung
Burggrafenstraße 4-10
Postfach 1107
1000 Berlin 30, FRG
Vocabulary standards:
Subject(s): copper; copper alloys
UDC 669. 3
Language(s): En; Fr; Ru

● *140* *(Commission/Committee)*
TC 27 "Solid mineral fuels", SC 1 "Coal preparation. Terminology and performance"/TC 27 "Combustibles minéraux solides", SC 1 "Préparation du charbon. Terminologie et rendement"
ISO/TC 27/SC 1
Parent organization: International Organization for Standardization
Contact: Standards Association of Australia (SAA)
Standards House
80–86 Arthur Street
North Sydney - N. S. W. 2060
Australia
Vocabulary standards:
Subject(s): solid mineral fuels
UDC 662. 62
Language(s): En; Fr; Ru

● *141* *(Commission/Committee)*
TC 27 "Solid mineral fuels", SC 1 "Coal preparation. Terminology and performance", WG 1 "Terminology"/TC 27 "Combustibles minéraux solides", SC 1 "Préparation du charbon. Terminologie et rendement", GT 1 "Terminologie"
ISO/TC 27/SC 1/WG 1
Parent organization: International Organization for Standardization
Contact: British Standards Institution (BSI)
2 Park Street
London W1A 2BS
UK

Vocabulary standards:
Subject(s): solid mineral fuels
UDC 662. 62
Language(s): En; Fr; Ru

● ***142*** *(Commission/Committee)*
TC 28 "Petroleum products and lubricants", SC 1 "Terminology"/TC 28 "Produits pétroliers et lubrifiants", SC 1 "Terminologie"
ISO/TC 28/SC 1
Parent organization: International Organization for Standardization
Contact: Association française de normalisation (AFNOR)
Tour Europe
Cedex 7
92080 Paris La Défense
France
Vocabulary standards:
Subject(s): petroleum products; lubricants
UDC 665. 7
Language(s): En; Fr; Ru

● ***143*** *(Commission/Committee)*
TC 28 "Petroleum products and lubricants", SC 1 "Terminology", WG 1 "Measurement terminology"/TC 28 "Produits pétroliers et lubrifiants", SC 1 "Terminologie", GT 1 "Terminologie du mesurage"
ISO/TC 28/SC 1/WG 1
Parent organization: International Organization for Standardization
Contact: British Standards Institution (BSI)
2 Park Street
London W1A 2BS
UK
Vocabulary standards:
Subject(s): petroleum products; lubricants
UDC 665. 7
Language(s): En; Fr; Ru

● ***144*** *(Commission/Committee)*
TC 28 "Petroleum products and lubricants", SC 6 "Bulk cargo transfer, accountability, inspection and reconciliation", WG 6 "Terminology"/TC 28 "Produits pétroliers et lubrifiants", SC 6 "Transfert des livraisons en vrac, prise en compte, inspection et résolution des divergences", GT 6 "Terminologie"
ISO/TC 28/SC 6/WG 6
Parent organization: International Organization for Standardization
Contact: British Standards Institution (BSI)
2 Park Street
London W1A 2BS
UK
Vocabulary standards:
Subject(s): petroleum products; lubricants
UDC 665. 7
Language(s): En; Fr; Ru

● ***145*** *(Commission/Committee)*
TC 30 "Measurement of fluid flow in closed conduits", SC 6 "Glossary of terms"/TC 30 "Mesure de débit des fluides dans les conduites fermées", SC 6 "Vocabulaire"
ISO/TC 30/SC 6
Parent organization: International Organization for Standardization
Contact: Association française de normalisation (AFNOR)
Tour Europe
Cedex 7
92080 Paris La Défense
France
Vocabulary standards:
Subject(s): fluid flow (measurement in closed conduits)
UDC 532. 57
Language(s): En; Fr; Ru

● ***146*** *(Commission/Committee)*
TC 31 "Tyres, rims and valves", WG 1 "Terminology"/ TC 31 "Pneus, jantes et valves", GT 1 "Terminologie"
ISO/TC 31/WG 1
Parent organization: International Organization for Standardization
Contact: Association française de normalisation (AFNOR)
Tour Europe
Cedex 7
92080 Paris La Défense
France
Vocabulary standards:
Subject(s): tyres; rims; valves
UDC 629. 11. 012. 5/. 6
Language(s): En; Fr; Ru

● ***147*** *(Commission/Committee)*
TC 33 "Refractories", SC 1 "Terminology"/TC 33 "Matériaux réfractaires" SC 1 "Terminologie"
ISO/TC 33/SC 1
Parent organization: International Organization for Standardization
Contact: Association française de normalisation (AFNOR)
Tour Europe
Cedex 7
92080 Paris La Défense
France
Vocabulary standards:
Subject(s): refractories
UDC 666. 76
Language(s): En; Fr; Ru

● ***148*** *(Commission/Committee)*
TC 34 "Agricultural food products", SC 4 "Cereals and pulses", WG 4 "Terminology"/TC 34 "Produits agricoles alimentaires", SC 4 "Céréales et légumineuses", GT 4 "Terminologie"
ISO/TC 34/SC 4/WG 4
Parent organization: International Organization for Standardization
Contact: Association française de normalisation (AFNOR)
Tour Europe
Cedex 7
92080 Paris La Défense
France
Vocabulary standards:
Subject(s): agricultural food products
UDC 633/637
Language(s): En; Fr; Ru

● ***149*** *(Commission/Committee)*
TC 34 "Agricultural food products", SC 14 "Fresh fruits and vegetables", WG 1 "Terminology"/TC 34 "Produits agricoles alimentaires", SC 14 "Fruits et légumes en l'état", GT 1 "Terminologie"
ISO/TC 34/SC 14/WG 1
Parent organization: International Organization for Standardization
Contact: Polish Committee for Standards
Measures and Quality Control (PKNiM)
Ul. Elektoralna 2
00–139 Warszawa
Poland
Vocabulary standards:
Subject(s): agricultural food products
UDC 633/637
Language(s): En; Fr; Ru

● ***150*** *(Commission/Committee)*
TC 34 "Agricultural food products", SC 15 "Coffee", WG 4 "Vocabulary"/TC 34 "Produits agricoles alimentaires", SC 15 "Café", GT 4 "Vocabulaire"
ISO/TC 34/SC 15/WG 4
Parent organization: International Organization for Standardization
Contact: Association française de normalisation (AFNOR)
Tour Europe
Cedex 7
92080 Paris La Défense
France
Vocabulary standards:
Subject(s): agricultural food products
UDC 633/637
Language(s): En; Fr; Ru

● ***151*** *(Commission/Committee)*
TC 35 "Paints and varnishes", SC 1 "Terminology"/ TC 35 "Peintures et vernis", SC 1 "Terminologie"
ISO/TC 35/SC 1
Parent organization: International Organization for Standardization
Contact: Association française de normalisation (AFNOR)
Tour Europe
Cedex 7
92080 Paris La Défense
France
Vocabulary standards:
Subject(s): paints and varnishes
UDC 667. 6
Language(s): En; Fr; Ru

● ***152*** *(Commission/Committee)*
TC 37 "Terminology (principles and co-ordination)"/ **TC 37 "Terminologie** (principes et coordination)"
ISO/TC 37
Parent organization: International Organization for Standardization
Contact: Mr. H. Felber
Österreichisches Normungsinstitut (ON)
Postfach 130
Heinestraße 38
A-1020 Wien
Austria
Vocabulary standards:
Subject(s): terminology (principles and methods); terminology (vocabulary)
UDC 001. 4. 001
Language(s): En; Fr
Theory: terminological principles; En; Fr
Publications: ISO/R 639–1967; ISO/R 704–1968; ISO/R 860–1968; ISO/R 919–1969; ISO/R 1087–1969; ISO/R 1149–1969; ISO 1951–1973

● ***153*** *(Commission/Committee)*
TC 37 "Terminology (principles and co-ordination)", **SC 1 "Principles of terminology"/TC 37 "Terminologie** (principes et coordination)", **SC 1 "Principes de terminologie"**
ISO/TC 37/SC 1
Parent organization: International Organization for Standardization
Contact: USSR State Committee for Standards (GOST)
Leninsky Prospekt 9
Moskva 117049
USSR
Vocabulary standards:
Subject(s): terminology (principles)
UDC 001. 4. 001
Language(s): En; Fr; Ru

● ***154*** *(Commission/Committee)*
TC 37 "Terminology (principles and co-ordination)", **SC 2 "Layout of vocabularies"/TC 37 "Terminologie** (principes et coordination)", **SC 2 "Présentation des vocabulaires"**
ISO/TC 37/SC 2
Parent organization: International Organization for Standardization
Contact: Standards Council of Canada (SCC).
Secretariat: Mr. P. Auger. Office de la langue française (OLF)
700 boul. St-Cyrille est
Québec
G1R 5G7
Canada
Vocabulary standards:
Subject(s): vocabularies (layout)
UDC 801. 316
Language(s): En; Fr; Ru

● ***155*** *(Commission/Committee)*
TC 37 "Terminology (principles and co-ordination)", **WG 4 "Computational aids in terminology and lexicography"/TC 37 "Terminologie** (principes et coordination)", **GT 4 "Aide à apporter par les ordinateurs dans les travaux de terminologie et de lexicographie"**
ISO/TC 37/WG 4
Parent organization: International Organization for Standardization
Contact: DIN Deutsches Institut für Normung
Burggrafenstraße 4–10
Postfach 1107
1000 Berlin 30, FRG
Vocabulary standards:
Subject(s): terminology (computational aids); lexicography (computational aids)
UDC 001. 4; UDC 681. 3; UDC 801. 316
Language(s): En; Fr; Ru

● ***156*** *(Commission/Committee)*
TC 37 "Terminology (principles and co-ordination)", **WG 5 "Vocabulary of terminology"/TC 37 "Terminologie** (principes et coordination)". **GT 5 "Vocabulaire de terminologie"**
ISO/TC 37/WG 5
Parent organization: International Organization for Standardization
Contact: Association française de normalisation (AFNOR)
Tour Europe
Cedex 7
92080 Paris la Défense
France
Committee members: 14
Established: 1981
Vocabulary standards:
Subject(s): terminology (vocabulary)
UDC 001. 4
Language(s): En; Fr
Publications: ISO/R 1087–1969 *Vocabulary of terminology*

● ***157*** *(Commission/Committee)*
TC 38 "Textiles", SC 11 "Care labelling of textiles and apparel", WG 2 "Terminology"/TC 38 "Textiles", SC 11 "Etiquetage d'entretien des textiles", "Terminologie"
ISO/TC 38/SC 11/WG 2
Parent organization: International Organization for Standardization
Contact: American National Standards Institute (ANSI)
1430 Broadway
New York
N. Y. 10018
USA

Vocabulary standards:
Subject(s): textiles
UDC 677
Language(s): En; Fr; Ru

● *158* *(Commission/Committee)*
TC 38 "Textiles", SC 12 "Textile floor coverings", WG 1 "Classification and terminology"/TC 38 "Textiles", SC 12 "Revêtements de sol textiles", GT 1 "Classification et terminologie"
ISO/TC 38/SC 12/WG 1
Parent organization: International Organization for Standardization
Contact: Association française de normalisation (AFNOR)
Tour Europe
Cedex 7
92080 Paris La Défense, France
Vocabulary standards:
Subject(s): textiles
UDC 677
Language(s): En; Fr; Ru

● *159* *(Commission/Committee)*
TC 38 "Textiles", SC 18 "Basic terminology, symbols, abbreviations and classification of fibres and yarns"/TC 38 "Textiles", SC 18 "Terminologie de base, symboles, abréviations et classification des fibres et des fils"
ISO/TC 38/SC 18
Parent organization: International Organization for Standardization
Contact: Ente Nazionale Italiano di Unificazione (UNI)
Piazza Armando Diaz 2
20123 Milano
Italy
Vocabulary standards:
Subject(s): textiles
UDC 677
Language(s): En; Fr; Ru

● *160* *(Commission/Committee)*
TC 38 "Textiles", SC 18 "Basic terminology, symbols, abbreviations and classification of fibres and yarns", WG 1 "Codification, abbreviations and symboles"/TC 38 "Textiles", SC 18 "Terminologie de base, symboles, abréviations et classification des fibres et des fils", GT 1 "Codification, abréviations et symboles"
ISO/TC 38/SC 18/WG 1
Parent organization: International Organization for Standardization
Contact: Vacant
Vocabulary standards:
Subject(s): textiles
UDC 677
Language(s): En; Fr; Ru

● *161* *(Commission/Committee)*
TC 38 "Textiles", SC 19 "Burning behaviour of textiles and textile products", WG 1 "Terminology"/ TC 38 "Textiles", SC 19 "Comportement au feu des textiles et des produits textiles", GT 1 "Terminologie"
ISO/TC 38/SC 19/WG 1
Parent organization: International Organization for Standardization
Contact: Standards Council of Canada (SCC)
International Standardization Branch
2000 Argentia Road
Suite 2-401
Mississauga
Ontario L5N 1VS
Canada
Vocabulary standards:
Subject(s): textiles
UDC 677
Language(s): En; Fr; Ru

● *162* *(Commission/Committee)*
TC 44 "Welding and allied processes", WG 1 "Definitions of weld positions on pipes"/TC 44 "Soudage et techniques connexes", GT 1 "Définitions des positions d'exécution des soudures sur tubes"
ISO/TC 44/WG 1
Parent organization: International Organization for Standardization
Contact: Nederlands Normalisatie-instituut (NNI)
Kalfjeslaan 2
P. O. Box 5059
2600 GB Delft
The Netherlands
Vocabulary standards:
Subject(s): welding (and allied processes)
UDC 621. 791
Language(s): En; Fr; Ru

● *163* *(Commission/Committee)*
TC 45 "Rubber and rubber products", WG 10 "Terminology"/ TC 45 "Elastomères et produits à base d'élastomères", GT 10 "Terminologie"
ISO/TC 45/WG 10
Parent organization: International Organization for Standardization
Contact: American National Standards Institute (ANSI)
1430 Broadway
New York
N. Y. 10018
USA
Vocabulary standards:
Subject(s): rubber; rubber products
UDC 678. 4; UDC 678. 7
Language(s): En; Fr; Ru

● *164* *(Commission/Committee)*
TC 46 "Documentation", SC 3 "Terminology of documentation"/TC 46 "Documentation", SC 3 "Terminologie de la documentation"
ISO/TC 46/SC 3
Parent organization: International Organization for Standardization
Contact: DIN Deutsches Institut für Normung
Burggrafenstraße 4–10
Postfach 1107
1000 Berlin 30
FRG
Vocabulary standards:
Subject(s): documentation
UDC 002
Language(s): En; Fr; Ru

● *165* *(Commission/Committee)*
TC 48 "Laboratory glassware and related apparatus", WG - "Terminology"/TC 48 "Verrerie de laboratoire et appareils connexes", GT - "Terminologie"
ISO/TC 48/WG -
Parent organization: International Organization for Standardization
Contact: British Standards Institution (BSI)
2 Park Street
London W1A 2BS
UK
Vocabulary standards:
Subject(s): laboratory glassware and related apparatus
UDC 542. 2
Language(s): En; Fr; Ru

● *166* *(Commission/Committee)*
TC 54 "Essential oils", WG 1 "Nomenclature"/TC 54 "Huiles essentielles", GT 1 "Nomenclature"
ISO/TC 54/WG 1
Parent organization: International Organization for Standardization

Contact: Nederlands Normalisatie-instituut (NNI)
Kalfjeslaan 2
P. O. Box 5059
2600 GB Delft
The Netherlands
Vocabulary standards:
Subject(s): essential oils
UDC 668. 5
Language(s): En; Fr; Ru

● ***167***(*Commission/Committee*)
TC 57 "Metrology and properties of surfaces", SC 3 "Instruments and devices for the measurement of roundness, straightness, cylindricity and flatness", WG 1 "Terminology"/TC 57 "Métrologie et propriétés des surfaces", SC 3 "Instruments et dispositifs de mesure de la circularité, de la rectitude, de la cylindricité et de la planéité", GT 1 "Terminologie"
ISO/TC 57/SC 3/WG 1
Parent organization: International Organization for Standardization
Contact: British Standards Institution (BSI)
2 Park Street
London W1A 2BS
UK
Vocabulary standards:
Subject(s): surfaces (metrology and properties)
UDC 620. 179. 11
Language(s): En; Fr; Ru

● ***168***(*Commission/Committee*)
TC 57 "Metrology and properties of surfaces", SC 4 "Geometrical parameters of the surface", WG 2 "Terminology"/TC 57 "Métrologie et propriétés des surfaces", SC 4 "Paramètres géométriques de la surface", GT 2 "Terminologie"
ISO/TC 57/SC 4/WG 2
Parent organization: International Organization for Standardization
Contact: DIN Deutsches Institut für Normung
Burggrafenstraße 4–10
Postfach 1107
1000 Berlin 30, FRG
Vocabulary standards:
Subject(s): surfaces (metrology and properties)
UDC 620. 179. 11
Language(s): En; Fr; Ru

● ***169***(*Commission/Committee*)
TC 59 "Building construction", SC 2 "Terminology, symbols and unification of language"/TC 59 "Construction immobilière", SC 2 "Terminologie, symboles et unification de langage"
ISO/TC 59/SC 2
Parent organization: International Organization for Standardization
Contact: Association française de normalisation (AFNOR)
Tour Europe
Cedex 7
92080 Paris La Défense
France
Vocabulary standards:
Subject(s): building construction
UDC 69
Language(s): En; Fr; Ru

● ***170***(*Commission/Committee*)
TC 59 "Building construction", SC 2 "Terminology, symbols and unification of language", WG 1 "Terminology"/TC 59 "Construction immobilière", SC 2 "Terminologie, symboles et unification de langage", GT 1 "Terminologie"
ISO/TC 59/SC 2/WG 1
Parent organization: International Organization for Standardization
Contact: Norges Standardiseringsforbund (NSF)
Postboks 7020 Homansbyen
Oslo 3
Norway
Vocabulary standards:
Subject(s): building construction
UDC 69
Language(s): En; Fr; Ru

● ***171***(*Commission/Committee*)
TC 60 "Gears", WG 4 "Terminology and notation of gears"/TC 60 "Engrenages", GT 4 "Terminologie et notation des engrenages"
ISO/TC 60/WG 4
Parent organization: International Organization for Standardization
Contact: Association française de normalisation (AFNOR)
Tour Europe
Cedex 7
92080 Paris La Défense
France
Vocabulary standards:
Subject(s): gears
UDC 621. 83
Language(s): En; Fr; Ru

● ***172***(*Commission/Committee*)
TC 61 "Plastics", SC 1 "Terminology"/TC 61 "Plastiques", SC 1 "Terminologie"
ISO/TC 61/SC 1
Parent organization: International Organization for Standardization
Contact: Association suisse de normalisation (SNV)
Kirchenweg 4
Postfach
8032 Zürich
Switzerland
Vocabulary standards:
Subject(s): plastics
UDC 678. 5/. 8
Language(s): En; Fr; Ru

● ***173***(*Commission/Committee*)
TC 61 "Plastics", SC 1 "Terminology", WG 1 "General vocabulary"/TC 61 "Plastiques", SC 1 "Terminologie", GT 1 "Vocabulaire général"
ISO/TC 61/SC 1/WG 1
Parent organization: International Organization for Standardization
Contact: American National Standards Institute (ANSI)
1430 Broadway
New York
N. Y. 10018
USA
Vocabulary standards:
Subject(s): plastics
UDC 678. 5/. 8
Language(s): En; Fr; Ru

● ***174***(*Commission/Committee*)
TC 61 "Plastics", SC 1 "Terminology", WG 2 "Special vocabularies"/TC 61 "Plastiques", SC 1 "Terminologie", GT 2 "Vocabulaires spéciaux"
ISO/TC 61/SC 1/WG 1
Parent organization: International Organization for Standardization
Contact: Instituto Nacional de Racionalización y Normalización (IRANOR)
Zurbano 46
Madrid 10
Spain

Vocabulary standards:
Subject(s): plastics
UDC 678. 5/. 8
Language(s): En; Fr; Ru

● *175**(Commission/Committee)*
TC 61 "Plastics", SC 4 "Burning behaviour", WG 4 "Nomenclature"/TC 61 "Plastiques", SC 4 "Comportement au feu", GT 4 "Nomenclature"
ISO/TC 61/SC 4/WG 4
Parent organization: International Organization for Standardization
Contact: Nederlands Normalisatie-instituut (NNI)
Kalfjeslaan 2
P. O. Box 5059
2600 GB Delft
The Netherlands
Vocabulary standards:
Subject(s): plastics
UDC 678. 5/. 8
Language(s): En; Fr; Ru

● *176**(Commission/Committee)*
TC 61 "Plastics", SC 1 "Terminology", WG 3 "Symbols"/TC 61 "Plastiques", SC 1 "Terminologie", GT 3 "Symboles"
ISO/TC 61/SC 1/WG 3
Parent organization: International Organization for Standardization
Contact: DIN Deutsches Institut für Normung
Burggrafenstraße 4–10
Postfach 1107
1000 Berlin 30, FRG
Vocabulary standards:
Subject(s): plastics
UDC 678. 5/. 8
Language(s): En; Fr; Ru

● *177**(Commission/Committee)*
TC 63 "Glass containers", SC 1 "Terminology"/TC 63 "Récipients en verre", SC 1 "Terminologie"
ISO/TC 63/SC 1
Parent organization: International Organization for Standardization
Contact: British Standards Institution (BSI)
2 Park Street
London W1A 2BS
UK
Vocabulary standards:
Subject(s): glass containers; containers
UDC 621. 798. 1; UDC 666. 17
Language(s): En; Fr; Ru

● *178**(Commission/Committee)*
TC 68 "Banking procedures", WG 2 "Vocabulary"/TC 68 "Procédures bancaires", GT 2 "Vocabulaire"
ISO/TC 68/WG 2
Parent organization: International Organization for Standardization
Contact: Association française de normalisation (AFNOR)
Tour Europe
Cedex 7
92080 Paris La Défense
France
Vocabulary standards:
Subject(s): banking procedures
UDC 336. 71
Language(s): En; Fr; Ru

● *179**(Commission/Committee)*
TC 69 "Applications of statistical methods", SC 1 "Terminology and symbols"/TC 69 "Application des méthodes statistiques", SC 1 "Terminologie et symboles"
ISO/TC 69/SC 1
Parent organization: International Organization for Standardization
Contact: Association française de normalisation (AFNOR)
Tour Europe
Cedex 7
92080 Paris La Défense
France
Vocabulary standards:
Subject(s): statistical methods
UDC 519. 2
Language(s): En; Fr; Ru

● *180**(Commission/Committee)*
TC 69 "Applications of statistical methods", SC 1 "Terminology and symbols", WG 1 "General statistical terms"/TC 69 "Application des méthodes statistiques", SC 1 "Terminologie et symboles", GT 1 "Termes statistiques généraux"
ISO/TC 69/SC 1/WG 1
Parent organization: International Organization for Standardization
Contact: Standards Council of Canada (SCC)
International Standardization Branch
2000 Argentia Road
Suite 2–401
Mississauga
Ontario L5N 1V8
Canada
Vocabulary standards:
Subject(s): statistical methods
UDC 519. 2
Language(s): En; Fr; Ru

● *181**(Commission/Committee)*
TC 69 "Applications of statistical methods", SC 1 "Terminology and symbols", WG 2 "Statistical quality control terms"/TC 69 "Application des méthodes statistiques", SC 1 "Terminologie et symboles", GT 2 "Termes relatifs au contrôle de statistique de qualité"
ISO/TC 69/SC 1/WG 2
Parent organization: International Organization for Standardization
Contact: British Standards Institution (BSI)
2 Park Street
London W1A 2BS
UK
Vocabulary standards:
Subject(s): statistical methods
UDC 519. 2
Language(s): En; Fr; Ru

● *182**(Commission/Committee)*
TC 70 "Internal combustion engines", SC 1 "Definitions"/TC 70 "Moteurs à combustion interne", SC 1 "Définitions"
ISO/TC 70/SC 1
Parent organization: International Organization for Standardization
Contact: USSR State Committee for Standards (GOST)
Leninsky Prospekt 9
Moskva 117049
USSR
Vocabulary standards:
Subject(s): combustion engines (internal)
UDC 621. 43
Language(s): En; Fr; Ru

● *183**(Commission/Committee)*
TC 71 "Concrete, reinforced concrete and pre-stressed concrete", SC 3 "Production and control of concrete", WG 3 "Terminology"/TC 71 "Béton, béton armé et béton précontraint", SC 3 "Fabrication, mise en oeuvre et contrôle du béton", GT 3 "Terminologie"

ISO/TC 71/SC 3/WG 3
Parent organization: International Organization for Standardization
Contact: British Standards Institution (BSI)
2 Park Street
London W1A 2BS
UK
Vocabulary standards:
Subject(s): concrete; reinforced concrete; pre-stressed concrete
UDC 691. 32
Language(s): En; Fr; Ru

● ***184*** *(Commission/Committee)*
TC 72 "Textile machinery, allied machinery, and accessories" WG 10 "Symbols for indicator plates on textile machinery"/TC 72 "Matériel pour l'industrie textile et matériel connexe", GT 10 "Symboles pour les plaques indicatrices sur les machines textiles"
ISO/TC 72/WG 10
Parent organization: International Organization for Standardization
Contact: Association française de normalisation (AFNOR)
Tour Europe
Cedex 7
92080 Paris La Défense
France
Vocabulary standards:
Subject(s): textile machinery
UDC 677. 05
Language(s): En; Fr; Ru

● ***185*** *(Commission/Committee)*
TC 73 "Consumer questions", SC 4 "Cutlery", WG 1 "Terminology"/TC 73 "Questions de consommation", SC 4 "Coutellerie de table", GT 1 "Terminologie"
ISO/TC 73/SC 4 /WG 1
Parent organization: International Organization for Standardization
Contact: Association française de normalisation (AFNOR)
Tour Europe
Cedex 7
92080 Paris La Défense
France
Vocabulary standards:
Subject(s): consumer questions
UDC 64
Language(s): En; Fr; Ru

● ***186*** *(Commission/Committee)*
TC 79 "Light metals and their alloys", SC 2 "Anodized aluminium", WG 5 "Terminology"/TC 79 "Métaux légers et leurs alliages", SC 2 "Aluminium anodisé", GT 5 "Terminologie"
ISO/TC 79/SC 2/WG 5
Parent organization: International Organization for Standardization
Contact: DIN Deutsches Institut für Normung
Burggrafenstraße 4–10
Postfach 1107
1000 Berlin 30, FRG
Vocabulary standards:
Subject(s): light metals (and alloys)
UDC 669. 71/. 72
Language(s): En; Fr; Ru

● ***187*** *(Commission/Committee)*
TC 79 "Light metals and their alloys", SC 9 "Symbolization"/TC 79 "Métaux légers et leurs alliages", SC 9 "Symbolisation"
ISO/TC 79/SC 9
Parent organization: International Organization for Standardization
Contact: Standards Council of Canada (SCC)
International Standardization Branch
2000 Argentia Road
Suite 2–401
Mississauga
Ontario L5N 1V8
Canada
Vocabulary standards:
Subject(s): light metals (and alloys)
UDC 669. 71/. 72
Language(s): En; Fr; Ru

● ***188*** *(Commission/Committee)*
TC 79 "Light metals and their alloys", WG 3 "Terminology"/TC 79 "Métaux légers et leurs alliages", GT 3 "Terminologie"
ISO/TC 79/WG 3
Parent organization: International Organization for Standardization
Contact: DIN Deutsches Institut für Normung
Burggrafenstraße 4–10
Postfach 1107
1000 Berlin 30, FRG
Vocabulary standards:
Subject(s): light metals (and alloys)
UDC 669. 71/. 72
Language(s): En; Fr; Ru

● ***189*** *(Commission/Committee)*
TC 81 "Common names for pesticides and other agrochemicals"/TC 81 "Noms communs pour les produits phytosanitaires et assimilés"
ISO/TC 81
Parent organization: International Organization for Standardization
Contact: British Standards Institution (BSI)
2 Park Street
London W1A 2BS
UK
Vocabulary standards:
Subject(s): pesticides; agrochemicals
UDC 632. 95
Language(s): En; Fr; Ru

● ***190*** *(Commission/Committee)*
TC 83 "Sports and recreational equipment", SC 4 "Skis", WG 1 "Terminology of skis"/TC 83 "Matériel de sport et d'activités de plein air", SC 4 "Skis", GT 1 "Terminologie des skis"
ISO/TC 83/SC 4/WG 1
Parent organization: International Organization for Standardization
Contact: DIN Deutsches Institut für Normung
Burggrafenstraße 4–10
Postfach 1107
1000 Berlin 30, FRG
Vocabulary standards:
Subject(s): sports equipment; recreational equipment
UDC 685. 6
Language(s): En; Fr; Ru

● ***191*** *(Commission/Committee)*
TC 83 "Sports and recreational equipment", SC 3 "Ski bindings", WG 1 "Terminology of ski bindings"/TC 83 "Matériel de sport et d'activités de plein air", SC 3 "Fixations de ski", GT 1 "Terminologie des fixations de ski"
ISO/TC 83/SC 3/WG 1
Parent organization: International Organization for Standardization
Contact: DIN Deutsches Institut für Normung
Burggrafenstraße 4–10
Postfach 1107
1000 Berlin 30, FRG

Vocabulary standards:
Subject(s): sports equipment; recreational equipment
UDC 685. 6
Language(s): En; Fr; Ru

● *192* *(Commission/Committee)*
TC 85 "Nuclear energy", SC 1 "Terminology, definitions, units and symbols"/TC 85 "Energie nucléaire", SC 1 "Terminologie, définitions, unités, symboles et schémas"
ISO/TC 85/SC 1
Parent organization: International Organization for Standardization
Contact: American National Standards Institute (ANSI)
1430 Broadway
New York
N. Y. 10018
USA
Vocabulary standards:
Subject(s): nuclear energy
UDC 539. 1
Language(s): En; Fr; Ru

● *193* *(Commission/Committee)*
TC 85 "Nuclear energy", SC 1 "Terminology, definitions, units and symbols", WG 1 "Definitions"/TC 85 "Energie nucléaire", SC 1 "Terminologie, définitions, unités, symboles et schémas", GT 1 "Définitions"
ISO/TC 85/SC 1/WG 1
Parent organization: International Organization for Standardization
Contact: American National Standards Institute (ANSI)
1430 Broadway
New York
N. Y. 10018
USA
Vocabulary standards:
Subject(s): nuclear energy
UDC 539. 1
Language(s): En; Fr; Ru

● *194* *(Commission/Committee)*
TC 92 "Fire tests on building materials, components and structures", WG 6 "Terminology"/TC 92 "Essais au feu sur les matériaux de construction, composants et structures", GT 6 "Terminologie"
ISO/TC 92/WG 6
Parent organization: International Organization for Standardization
Contact: Association française de normalisation (AFNOR)
Tour Europe
Cedex 7
92080 Paris La Défense
France
Vocabulary standards:
Subject(s): fire tests (on building materials, components and structures)
UDC 699. 81; UDC 620. 1
Language(s): En; Fr; Ru

● *195* *(Commission/Committee)*
TC 93 "Starch (including derivatives and by-products)", **SC 1 "General problems", WG 1 "Terminology"/TC 93 "Amidon** (amidons, fécules), **dérivés et sous-produits", SC 1 "Problèmes généraux", GT 1 "Terminologie"**
ISO/TC 93/SC 1/WG 1
Parent organization: International Organization for Standardization
Contact: Association française de normalisation (AFNOR)
Tour Europe
Cedex 7
92080 Paris La Défense
France
Vocabulary standards:
Subject(s): starch (including derivatives and by-products)
UDC 664. 2
Language(s): En; Fr; Ru

● *196* *(Commission/Committee)*
TC 94 "Personal safety - Protective clothing and equipment", SC 9 "Protective clothing against heat and fire", WG 3 "Definitions"/TC 94 "Sécurité individuelle - Vêtements et équipments de protection", SC 9 "Vêtements de protection contre la chaleur et le feu", GT 3 "Définitions"
ISO/TC 94/SC 9/WG 3
Parent organization: International Organization for Standardization
Contact: Association française de normalisation (AFNOR)
Tour Europe
Cedex 7
92080 Paris La Défense
France
Vocabulary standards:
Subject(s): personal safety; protective clothing (against heat and fire)
UDC 614. 89; UDC 687
Language(s): En; Fr; Ru

● *197* *(Commission/Committee)*
TC 95 "Office machines", SC 4 "Duplicating and document copying machines", WG 2 "Terminology of duplicating and copying machines"/TC 95 "Machines de bureau", SC 4 "Duplicateurs et machines à reproduire", GT 2 "Terminologie des duplicateurs et des machines à reproduire"
ISO/TC 95/SC 4/WG 2
Parent organization: International Organization for Standardization
Contact: British Standards Institution (BSI)
2 Park Street
London W1A 2BS
UK
Vocabulary standards:
Subject(s): office machines
UDC 681. 6
Language(s): En; Fr; Ru

● *198* *(Commission/Committee)*
TC 96 "Cranes; lifting appliances and related excavator equipment" (provisional), **SC 2 "Terminology"/TC 96 "Grues, appareils de levage et équipements de pelles correspondants" (provisoire), SC 2 "Terminologie"**
ISO/TC 96/SC 2
Parent organization: International Organization for Standardization
Contact: USSR State Committee for Standards (GOST)
Leninsky Prospekt 9
Moskva 117049
USSR
Vocabulary standards:
Subject(s): cranes; lifting appliances; excavator equipment
UDC 621. 873
Language(s): En; Fr; Ru

● *199* *(Commission/Committee)*
TC 97 "Computers and information processing", SC 1 "Vocabulary"/TC 97 "Calculateurs et traitement de l'information", SC 1 "Vocabulaire"
ISO/TC 97/SC 1
Parent organization: International Organization for Standardization
Contact: Association française de normalisation (AFNOR)

Tour Europe
Cedex 7
92080 Paris La Défense
France
Vocabulary standards:
Subject(s): computers; information processing
UDC 681. 3
Language(s): En; Fr; Ru

● ***200***(*Commission/Committee*)
TC 97 "Computers and information processing", SC 1 "Vocabulary", WG 1 "Maintenance"/TC 97 "Calculateurs et traitement de l'information", SC 1 "Vocabulaire", GT 1 "Maintenance"
ISO/TC 97/SC 1/WG 1
Parent organization: International Organization for Standardization
Contact: Association française de normalisation (AFNOR)
Tour Europe
Cedex 7
92080 Paris La Défense
France
Vocabulary standards:
Subject(s): computers; information processing
UDC 681. 3
Language(s): En; Fr; Ru

● ***201***(*Commission/Committee*)
TC 97 "Computers and information processing", SC 1 "Vocabulary", WG 2 "Editing"/TC 97 "Calculateurs et traitement de l'information", SC 1 "Vocabulaire", GT 2 "Rédaction"
ISO/TC 97/SC 1/WG 2
Parent organization: International Organization for Standardization
Contact: Association française de normalisation (AFNOR)
Tour Europe
Cedex 7
92080 Paris La Défense
France
Vocabulary standards:
Subject(s): computers; information processing
UDC 681. 3
Language(s): En; Fr; Ru

● ***202***(*Commission/Committee*)
TC 97 "Computers and information processing", SC 18 "Text preparation and interchange", WG 2 "Symbols and terminology"/TC 97 "Calculateurs et traitement de l'information", SC 18 "Préparation et échange de textes", GT 2 "Symboles et terminologie"
ISO/TC 97/SC 18/WG 2
Parent organization: International Organization for Standardization
Contact: Japanese Industrial Standards Committee (JISC)
c/o Standards Department
Agency of Industrial Science and Technology
Ministry of International Trade and Industry
1-3-1 Kasumigaseki
Chiyodaku
Tokyo 100
Japan
Vocabulary standards:
Subject(s): computers; information processing
UDC 681. 3
Language(s): En; Fr; Ru

● ***203***(*Commission/Committee*)
TC 98 "Bases for design of structures", SC 1 "Terminology and symbols"/TC 98 "Bases du calcul des constructions", SC 1 "Terminologie et symboles"
ISO/TC 98/SC 1
Parent organization: International Organization for Standardization
Contact: Association française de normalisation (AFNOR)
Tour Europe
Cedex 7
92080 Paris La Défense
France
Vocabulary standards:
Subject(s): structures (design)
UDC 624. 04
Language(s): En; Fr; Ru

● ***204***(*Commission/Committee*)
TC 99 "Semi-manufactures of timber", WG 1 "Terminology"/TC 99 "Bois semi-manufacturés", GT 1 "Terminologie"
ISO/TC 99/WG 1
Parent organization: International Organization for Standardization
Contact: Vacant
Vocabulary standards:
Subject(s): timber
UDC 674. 06
Language(s): En; Fr; Ru

● ***205***(*Commission/Committee*)
TC 101 "Continuous mechanical handling equipment", WG 1 "Terminology"/TC 101 "Engins de manutention continue", GT 1 "Terminologie"
ISO/TC 101/WG 1
Parent organization: International Organization for Standardization
Contact: Association française de normalisation (AFNOR)
Tour Europe
Cedex 7
92080 Paris La Défense
France
Vocabulary standards:
Subject(s): handling equipment
UDC 621. 867
Language(s): En; Fr; Ru

● ***206***(*Commission/Committee*)
TC 105 "Steel wire ropes", SC 1 "Terminology, definitions and symbols"/TC 105 "Câbles en acier", SC 1 "Terminologie, définitions et symboles"
ISO/TC 105/SC 1
Parent organization: International Organization for Standardization
Contact: Vacant
Vocabulary standards:
Subject(s): ropes; steel wire ropes
UDC 677. 721
Language(s): En; Fr; Ru

● ***207***(*Commission/Committee*)
TC 106 "Dentistry", WG 3 "Terminology"/TC 106 "Produits et matériel pour l'art dentaire", GT 3 "Terminologie"
ISO/TC 106/WG 3
Parent organization: International Organization for Standardization
Contact: Association française de normalisation (AFNOR)
Tour Europe
Cedex 7
92080 Paris La Défense
France
Vocabulary standards:
Subject(s): dentistry
UDC 616. 314
Language(s): En; Fr; Ru

● ***208***(*Commission/Committee*)
TC 107 "Metallic and other non-organic coatings", SC 1

"Terminology"/TC 107 "Revêtements métalli ques et autres revêtements non organiques", SC 1 "Terminologie"
ISO/TC 107/SC 1
Parent organization: International Organization for Standardization
Contact: Association suisse de normalisation (SNV)
Kichenweg 4
Postfach
8032 Zürich
Switzerland
Vocabulary standards:
Subject(s): coatings
UDC 621. 793/. 795
Language(s): En; Fr; Ru

● ***209*** *(Commission/Committee)*
TC 108 "Mechanical vibration and shock", SC 1 "Balancing, including balancing machines", WG 1 "Balancing terminology"/TC 108 "Vibrations et chocs mécaniques", SC 1 "Equilibrage, y compris les machines à équilibrer", GT 1 "Terminologie d'équilibrage"
ISO/TC 108/SC 1/WG 1
Parent organization: International Organization for Standardization
Contact: British Standards Institution (BSI)
2 Park Street
London W1A 2BS
U. K.
Vocabulary standards:
Subject(s): mechanical vibration; mechanical shock
UDC 534. 1/. 5; UDC 621-752
Language(s): En; Fr; Ru

● ***210*** *(Commission/Committee)*
TC 108 "Mechanical vibration and shock", SC 4 "Human exposure to mechanical vibration and shock", WG 1 "Terminology"/TC 108 "Vibrations et chocs mécaniques", SC 4 "Exposition des individus aux vibrations et chocs mécaniques", GT 1 "Terminologie"
ISO/TC 108/SC 4/WG 1
Parent organization: International Organization for Standardization
Contact: American National Standards Institute (ANSI)
1430 Broadway
New York
N. Y. 10018
USA
Vocabulary standards:
Subject(s): mechanical vibration; mechanical shock
UDC 534. 1/. 5; UDC 621-752
Language(s): En; Fr; Ru

● ***211*** *(Commission/Committee)*
TC 108 "Mechanical vibration and shock", WG 1 "Terminology"/TC 108 "Vibrations et chocs mé caniques", GT 1 "Terminologie"
ISO/TC 108/WG 1
Parent organization: International Organization for Standardization
Contact: American National Standards Institute (ANSI)
1430 Broadway
New York
N. Y. 10018
USA
Vocabulary standards:
Subject(s): mechanical vibration; mechanical shock
UDC 534. 1/. 5; UDC 621-752
Language(s): En; Fr; Ru

● ***212*** *(Commission/Committee)*
TC 109 "Oil burners and associated equipment", SC 1 "Terminology"/TC 109 "Brûleurs à combustible liquide et leur équipement", SC 1 "Terminologie"
ISO/TC 109/SC 1
Parent organization: International Organization for Standardization
Contact: Vacant
Vocabulary standards:
Subject(s): oil burners (and associated equipment)
UDC 662. 94
Language(s): En; Fr; Ru

● ***213*** *(Commission/Committee)*
TC 110 "Industrial trucks", SC 1 "General terminology"/ TC 110 "Chariots de manutention", SC 1 "Terminologie générale"
ISO/TC 110/SC 1
Parent organization: International Organization for Standardization
Contact: Association française de normalisation (AFNOR)
Tour Europe
Cedex 7
92080 Paris La Défense
France
Vocabulary standards:
Subject(s): trucks
UDC 621. 868. 2
Language(s): En; Fr; Ru

● ***214*** *(Commission/Committee)*
TC 112 "Vacuum technology", SC 4 "Terminology"/ TC 112 "Technique du vide", SC 4 "Terminologie"
ISO/TC 112/SC 4
Parent organization: International Organization for Standardization
Contact: Institut belge de normalisation (IBN)
Av. de la Brabançonne
29
1040 Bruxelles
Belgium
Vocabulary standards:
Subject(s): vacuum technology
UDC 621. 52
Language(s): En; Fr; Ru

● ***215*** *(Commission/Committee)*
TC 113 "Measurement of liquid flow in open channels", SC 3 "Glossary of terms"/TC 113 "Mesure de débit des liquides dans les canaux découverts" SC 3 "Vocabulaire des termes"
ISO/TC 113/SC 3
Parent organization: International Organization for Standardization
Contact: India Standards Institution (ISI)
Manak Bhavan
9 Bahadur Shah Zafar Marg
New Delhi 110002
India
Vocabulary standards:
Subject(s): liquid flow (measurement in open channels)
UDC 532. 57; UDC 532. 543; UDC 627. 133
Language(s): En; Fr; Ru

● ***216*** *(Commission/Committee)*
TC 114 "Horology", SC 8 "Definitions of the chronometer"/ TC 114 "Horlogerie", SC 8 "Définitions du chronomètre"
ISO/TC 114/SC 8
Parent organization: International Organization for Standardization
Contact: Association suisse de normalisation (SNV)
Kirchenweg 4
Postfach
8032 Zürich
Switzerland
Vocabulary standards:
Subject(s): horology

UDC 681. 11
Language(s): En; Fr; Ru

● ***217*** *(Commission/Committee)*
TC 114 "Horology", SC 9 "Technical definitions"/ TC 114 "Horlogerie", SC 9 "Définitions techniques"
ISO/TC 114/SC 9
Parent organization: International Organization for Standardization
Contact: Association française de normalisation (AFNOR)
Tour Europe
Cedex 7
92080 Paris La Défense
France
Vocabulary standards:
Subject(s): horology
UDC 681. 11
Language(s): En; Fr; Ru

● ***218*** *(Commission/Committee)*
TC 114 "Horology", SC 9 "Technical definitions", WG 1 "Preparation of technical definitions"/TC 114 "Horlogerie", SC 9 "Définitions techniques", GT 1 "Elaboration des définitions techniques"
ISO/TC 114/SC 9/WG 1
Parent organization: International Organization for Standardization
Contact: Association française de normalisation (AFNOR)
Tour Europe
Cedex 7
92080 Paris La Défense
France
Vocabulary standards:
Subject(s): horology
UDC 681. 11
Language(s): En; Fr; Ru

● ***219*** *(Commission/Committee)*
TC 114 "Horology", SC 4 "Terminology"/TC 114 "Horlogerie", SC 4 "Terminologie"
ISO/TC 114/SC 4
Parent organization: International Organization for Standardization
Contact: Association suisse de normalisation (SNV)
Kirchenweg 4
Postfach
8032 Zürich
Switzerland
Vocabulary standards:
Subject(s): horology
UDC 681. 11
Language(s): En; Fr; Ru

● ***220*** *(Commission/Committee)*
TC 116 "Space heating appliance", SC 1 "Terminology"/ TC 116 "Appareils de chauffage", SC 1 "Terminologie"
ISO/TC 116/SC 1
Parent organization: International Organization for Standardization
Contact: DIN Deutsches Institut für Normung
Burggrafenstraße 4–10
Postfach 1107
1000 Berlin 30, FRG
Vocabulary standards:
Subject(s): space heating appliance
UDC 697. 1/. 8
Language(s): En; Fr; Ru

● ***221*** *(Commission/Committee)*
TC 118 "Compressors, pneumatic tools and pneumatic machines", SC 2 "Terminology"/TC 118 "Compresseurs, outils et machines pneumatiques", SC 2 "Terminologie"
ISO/TC 118/SC 2
Parent organization: International Organization for Standardization
Contact: Vacant
Vocabulary standards:
Subject(s): compressors; pneumatic tools; pneumatic machines
UDC 621. 51. /54
Language(s): En; Fr; Ru

● ***222*** *(Commission/Committee)*
TC 119 "Powder metallurgy", SC 1 "Terminology"/ TC 119 "Métallurgie des poudres", SC 1 "Terminologie"
ISO/TC 119/SC 1
Parent organization: International Organization for Standardization
Contact: British Standards Institution (BSI)
2 Park Street
London W1A 2BS
UK
Vocabulary standards:
Subject(s): powder metallurgy
UDC 621. 762
Language(s): En; Fr; Ru

● ***223*** *(Commission/Committee)*
TC 119 "Powder metallurgy", SC 5 "Specifications for powder metallurgical materials (excluding hardmetals)**", WG 1 "Principles of nomenclature and designation"/TC 119 "Métallurgie des poudres", SC 5 "Spécifications pour les matériaux de la métallurgie des poudres** (à l'exclusion des métaux-durs)**", GT 1 "Principes de nomenclature et de désignation"**
ISO/TC 119/SC 5/WG 1
Parent organization: International Organization for Standardization
Contact: SIS - Standardiseringskommissionen i Sverige
Tegnérgatan 11
Box 3 295
103 66 Stockholm
Sweden
Vocabulary standards:
Subject(s): powder metallury
UDC 621. 762
Language(s): En; Fr; Ru

● ***224*** *(Commission/Committee)*
TC 120 "Leather", WG 1 "Terminology"/TC 120 "Cuir", GT 1 "Terminologie"
ISO/TC 120/WG 1
Parent organization: International Organization for Standardization
Contact: Vacant
Vocabulary standards:
Subject(s): leather
UDC 675
Language(s): En; Fr; Ru

● ***225*** *(Commission/Committee)*
TC 121 "Anaesthetic equipment and medical breathing machines", SC 4 "Anaesthesia terminology"/TC 121 "Matériel d'anesthésie et respirateurs médicaux", SC 4 "Terminologie concernant l'anesthésie"
ISO/TC 121/SC 4
Parent organization: International Organization for Standardization
Contact: Association française de normalisation (AFNOR)
Tour Europe
Cedex 7
92080 Paris La Défense
France

Vocabulary standards:
Subject(s): anaesthetic equipment; medical breathing machines
UDC 616–089. 5
Language(s): En; Fr; Ru

● ***226*** *(Commission/Committee)*
TC 122 "Packaging", SC 4 "Terminology, codification, classification and symbols"/TC 122 "Emballages", SC 4 "Terminologie, codification, classification et symboles"
ISO/TC 122/SC 4
Parent organization: International Organization for Standardization
Contact: DIN Deutsches Institut für Normung
Burggrafenstraße 4–10
Postfach 1107
1000 Berlin 30, FRG
Vocabulary standards:
Subject(s): packaging
UDC 621. 798
Language(s): En; Fr; Ru

● ***227*** *(Commission/Committee)*
TC 122 "Packaging", SC 4 "Terminology, codification, classification and symbols", WG 1 "Classification and terminology of packaging"/TC 122 "Emballages", SC 4 "Terminologie, codification, classification et symboles", GT 1 "Classification et terminologie de l'emballage"
ISO/TC 122/SC 4/WG 1
Parent organization: International Organization for Standardization
Contact: Association française de normalisation (AFNOR)
Tour Europe
Cedex 7
92080 Paris La Défense
France
Vocabulary standards:
Subject(s): packaging
UDC 621. 798
Language(s): En; Fr; Ru

● ***228*** *(Commission/Committee)*
TC 123 "Plain bearings", SC 1 "Terms, definitions and classifications"/TC 123 "Paliers lisses", SC 1 "Terminologie, définitions et classification"
ISO/TC 123/SC 1
Parent organization: International Organization for Standardization
Contact: USSR State Committee of Standards (GOST)
Leninsky Prospekt 9
Moskva 117049
USSR
Vocabulary standards:
Subject(s): bearings
UDC 621. 822. 5
Language(s): En; Fr; Ru

● ***229*** *(Commission/Committee)*
TC 125 "Enclosures and conditions for testing", WG 1 "Terminology"/TC 125 "Enceintes et conditions d'essai", GT 1 "Terminologie"
ISO/TC 125/WG 1
Parent organization: International Organization for Standardization
Contact: Association française de normalisation (AFNOR) (provisional)
Tour Europe
Cedex 7
92080 Paris La Défense
France
Vocabulary standards:
Subject(s): testing (enclosures and conditions for)
UDC 620. 1. 05
Language(s): En; Fr; Ru

● ***230*** *(Commission/Committee)*
TC 130 "Graphic technology", SC 1 "Terminology"/TC 130 "Technologie graphique", SC 1 "Terminologie"
ISO/TC 130/SC 1
Parent organization: International Organization for Standardization
Contact: Association française de normalisation (AFNOR)
Tour Europe
Cedex 7
92080 Paris La Défense
France
Vocabulary standards:
Subject(s): graphic technology
UDC 655. 1/.3; UDC 681. 6
Language(s): En; Fr; Ru

● ***231*** *(Commission/Committee)*
TC 131 "Fluid power systems and components", SC 1 "Terminology, classification and symbols"/TC 131 "Transmissions hydrauliques et pneumatiques", SC 1 "Terminologie, classification et symboles"
ISO/TC 131/SC 1
Parent organization: International Organization for Standardization
Contact: Japanese Industrial Standards Committee (JISC)
c/o Standards Department
Agency of Industrial Science and Technology
Ministry of International Trade and Industry
1-3-1 Kasumigaseki
Chiyodaku
Tokyo 100
Japan
Vocabulary standards:
Subject(s): fluid power systems
UDC 621. 22; UDC 621. 5
Language(s): En; Fr; Ru

● ***232*** *(Commission/Committee)*
TC 131 "Fluid power systems and components", SC 1 "Terminology, classification and symbols", WG 1 "Graphic symbols"/TC 131 "Transmissions hydrauliques et pneumatiques", SC 1 "Terminologie, classification et symboles", GT 1 "Symboles graphiques"
ISO/TC 131/SC 1/WG 1
Parent organization: International Organization for Standardization
Contact: Nederlands Normalisatie-instituut (NNI)
Kalfjeslaan 2
P. O. Box 5059
2600 GB Delft
The Netherlands
Vocabulary standards:
Subject(s): fluid power systems
UDC 621. 22; UDC 621. 5
Language(s): En; Fr; Ru

● ***233*** *(Commission/Committee)*
TC 131 "Fluid power systems and components", SC 1 "Terminology, classification and symbols", WG 2 "Terminology"/TC 131 "Transmissions hydrauliques et pneumatiques", SC 1 "Terminologie, classification et symboles", GT 2 "Terminologie"
ISO/TC 131/SC 1/WG 2
Parent organization: International Organization for Standardization
Contact: Japanese Industrial Standards Committee (JISC)
c/o Standards Department
Agency of Industrial Science and Technology
Ministry of International Trade and Industry
1-3-1 Kasumigaseki

Chiyodaku
Tokyo 100
Japan
Vocabulary standards:
Subject(s): fluid power systems
UDC 621. 22; UDC 621. 5
Language(s): En; Fr; Ru

● ***234*** *(Commission/Committee)*
TC 131 "Fluid power systems and components", SC 7 "Sealing devices", WG 5 "Terminology of rotary shaft lip type seals"/ TC 131 "Transmissions hydrauliques et pneumatiques", SC 7 "Dispositifs d'étanchéité", GT 5 "Terminologie des bagues d'étanchéité à lèvres pour arbres rotatifs"
ISO/TC 131/SC 7/WG 5
Parent organization: International Organization for Standardization
Contact: Japanese Industrial Standards Committee (JISC)
c/o Standards Department
Agency of Industrial Science and Technology
Ministry of International Trade and Industry
1–3–1 Kasumigaseki
Chiyodaku
Tokyo 100
Japan
Vocabulary standards:
Subject(s): fluid power systems
UDC 621. 22; UDC 621. 5
Language(s): En; Fr; Ru

● ***235*** *(Commission/Committee)*
TC 134 "Fertilizers and soil conditioners", SC 1 "Terminology and labelling"/TC 134 "Fertilisants", SC 1 "Terminologie et étiquetage"
ISO/TC 134/SC 1
Parent organization: International Organization for Standardization
Contact: Association française de normalisation (AFNOR)
Tour Europe
Cedex 7
92080 Paris La Défense
France
Vocabulary standards:
Subject(s): fertilizers; soil conditioners
UDC 631. 8
Language(s): En; Fr; Ru

● ***236*** *(Commission/Committee)*
TC 134 "Fertilizers and soil conditioners", SC 1 "Terminology and labelling", WG 1 "Glossary"/TC 134 "Fertilisants", SC 1 "Terminologie et étiquetage", GT 1 "Glossaire"
ISO/TC 134/SC 1/WG 1
Parent organization: International Organization for Standardization
Contact: Association française de normalisation (AFNOR)
Tour Europe
Cedex 7
92080 Paris La Défense
France
Vocabulary standards:
Subject(s): fertilizers; soil conditioners
UDC 631. 8
Language(s): En; Fr; Ru

● ***237*** *(Commission/Committee)*
TC 135 "Non-destructive testing", SC 1 "Terminology"/TC 135 "Essais non destructifs", SC 1 "Terminologie"
ISO/TC 135/SC 1
Parent organization: International Organization for Standardization
Contact: Association française de normalisation (AFNOR)
Tour Europe
Cedex 7
92080 Paris La Défense
France
Vocabulary standards:
Subject(s): non-destructive testing
UDC 620. 179. 1
Language(s): En; Fr; Ru

● ***238*** *(Commission/Committee)*
TC 136 'Furniture", SC 4 "Terms and definitions"/ TC 136 "Ameublement", SC 4 "Termes et définitions"
ISO/TC 136/SC 4
Parent organization: International Organization for Standardization
Contact: Vacant
Vocabulary standards:
Subject(s): furniture
UDC 645. 4 : 684. 4
Language(s): En; Fr; Ru

● ***239*** *(Commission/Committee)*
TC 136 "Furniture", SC 9 "Hardware for furniture", WG 1 "General classification and general terminology"/TC 136 "Ameublement", SC 9 "Quincaillerie d'ameublement", GT 1 "Classification générale et terminologie générale"
ISO/TC 136/SC 9/WG 1
Parent organization: International Organization for Standardization
Contact: Österreichisches Normungsinstitut (ON)
Heinestr. 38
Postfach 130
1021 Wien
Austria
Vocabulary standards:
Subject(s): furniture
UDC 645. 4 : 684. 4
Language(s): En; Fr; Ru

● ***240*** *(Commission/Committee)*
TC 142 "Cleaning equipment for air and other gases", WG 1 "Terminology"/TC 142 "Séparateurs aérauliques", GT 1 "Terminologie"
ISO/TC 142/WG 1
Parent organization: International Organization for Standardization
Contact: Association française de normalisation (AFNOR)
Tour Europe
Cedex 7
92080 Paris La Défense
France
Vocabulary standards:
Subject(s): cleaning equipment (for air and other gases)
UDC 66. 074
Language(s): En; Fr; Ru

● ***241*** *(Commission/Committee)*
TC 146 "Air quality", SC 4 "General aspects", WG 1 "Terminology"/TC 146 "Qualité de l'air", SC 4 "Aspects généraux", GT 1 "Terminologie"
ISO/TC 146/SC 4/WG 1
Parent organization: International Organization for Standardization
Contact: British Standards Institution (BSI)
2 Park Street
London W1A 2BS
UK
Vocabulary standards:
Subject(s): air quality
UDC 614. 71/. 76
Language(s): En; Fr; Ru

● *242* *(Commission/Committee)*
TC 147 "Water quality", SC 1 "Terminology"/TC 147 "Qualité de l'eau", SC 1 "Terminologie"
ISO/TC 147/SC 1
Parent organization: International Organization for Standardization
Contact: South African Bureau of Standards (SABS)
Private Bag X 191
Pretoria 0001
South Africa
Vocabulary standards:
Subject(s): water quality
UDC 628. 1/. 3
Language(s): En; Fr; Ru

● *243* *(Commission/Committee)*
TC 147 "Water quality", SC 1 "Terminology", WG 1 "Definitions"/TC 147 "Qualité de l'eau", SC 1 "Terminologie", GT 1 "Définitions"
ISO/TC 147/SC 1/WG 1
Parent organization: International Organization for Standardization
Contact: Standards Council of Canada (SCC)
International Standardization Branch
2000 Argentia Road
Suite 2–401
Mississauga
Ontario
L5N 1V8
Canada
Vocabulary standards:
Subject(s): water quality
UDC 628. 1/. 3
Language(s): En; Fr; Ru

● *244* *(Commission/Committee)*
TC 148 "Sewing machines", WG 1 "Classification and terminology"/TC 148 "Machines à coudre", GT 1 "Classification et terminologie"
ISO/TC 148/WG 1
Parent organization: International Organization for Standardization
Contact: Association française de normalisation (AFNOR)
Tour Europe
Cedex 7
92080 Paris La Défense
France
Vocabulary standards:
Subject(s): sewing machines
UDC 687. 053
Language(s): En; Fr; Ru

● *245* *(Commission/Committee)*
TC 149 "Cycles", WG 1 "Cycle terminology"/TC 149 "Cycles", GT 1 "Terminologie des cycles"
ISO/TC 149/WG 1
Parent organization: International Organization for Standardization
Contact: Association française de normalisation (AFNOR)
Tour Europe
Cedex 7
92080 Paris La Défense
France
Vocabulary standards:
Subject(s): cycles
UDC 629. 118
Language(s): En; Fr; Ru

● *246* *(Commission/Committee)*
TC 150 "Implants for surgery", WG 1 "Terminology"/ TC 150 "Implants pour la chirurgie", GT 1 "Terminologie"
ISO/TC 150/WG 1
Parent organization: International Organization for Standardization
Contact: British Standards Institution (BSI)
2 Park Street
London W1A 2BS
UK
Vocabulary standards:
Subject(s): surgery (implants for)
UDC 615. 46
Language(s): En; Fr; Ru

● *247* *(Commission/Committee)*
TC 154 "Documents and data elements in administration, commerce and industry", SC 1 "Terminology"/ TC 154 "Documents et éléments d'information dans l'administration, le commerce et l'industrie", SC 1 "Terminologie"
ISO/TC 154/SC 1
Parent organization: International Organization for Standardization
Contact: Standards Council of Canada (SCC)
International Standardization Branch
2000 Argentia Road
Suite 2–401
Mississauga
Ontario
L5N 1V8
Canada
Vocabulary standards:
Subject(s): administration; commerce; industry
UDC 651. 5
Language(s): En; Fr; Ru

● *248* *(Commission/Committee)*
TC 155 "Nickel and nickel alloys", WG 1 "Terminology"/ TC 155 "Nickel et alliages de nickel", GT 1 "Terminologie"
ISO/TC 155/WG 1
Parent organization: International Organization for Standardization
Contact: DIN Deutsches Institut für Normung
Burggrafenstraße 4–10
Postfach 1107
1000 Berlin 30, FRG
Vocabulary standards:
Subject(s): nickel ; nickel alloys
UDC 669. 24
Language(s): En; Fr; Ru

● *249* *(Commission/Committee)*
TC 156 "Corrosion of metals and alloys", WG 1 "Terminology"/TC 156 "Corrosion des métaux et alliages", GT 1 "Terminologie"
ISO/TC 156/WG 1
Parent organization: International Organization for Standardization
Contact: USSR State Committee for Standards (GOST)
Leninsky Prospekt 9
Moskva 117049
USSR
Vocabulary standards:
Subject(s): corrosion (of metals and alloys)
UDC 620. 193 : 669. 018. 8
Language(s): En; Fr; Ru

● *250* *(Commission/Committee)*
TC 158 "Analysis of gases", WG 1 "Terminology"/ TC 158 "Analyse des gaz", GT 1 "Terminologie"
ISO/TC 158/WG 1
Parent organization: International Organization for Standardization
Contact: American National Standards Institute (ANSI)
1430 Broadway
New York
N. Y. 10018
USA

Vocabulary standards:
Subject(s): gases (analysis of)
UDC 66. 07
Language(s): En; Fr; Ru

● ***251*** *(Commission/Committee)*
TC 162 "Doors and windows", SC 3 "Hardware", WG 1 "Terminology for locks and lock parts for doors"/ TC 162 "Portes et fenêtres", SC 3 "Quincaillerie", GT 1 "Terminologie des serrures de portes et des éléments de ces serrures"
ISO/TC 162/SC 3/WG 1
Parent organization: International Organization for Standardization
Contact: SIS - Standardiseringskommissionen i Sverige
Tegnérgatan 11
103 66 Stockholm
Sweden
Vocabulary standards:
Subject(s): doors; windows
UDC 69. 028
Language(s): En; Fr; Ru

● ***252*** *(Commission/Committee)*
TC 162 "Doors and windows", SC 3 "Hardware", WG 2 "Terminology for automatic door closers"/TC 162 "Portes et fenêtres", SC 3 "Quincaillerie", GT 2 "Terminologie des ferme-portes automatiques"
ISO/TC 162/SC 3/WG 2
Parent organization: International Organization for Standardization
Contact: British Standards Institution (BSI)
2 Park Street
London W1A 2BS
UK
Vocabulary standards:
Subject(s): doors; windows
UDC 69. 028
Language(s): En; Fr; Ru

● ***253*** *(Commission/Committee)*
TC 162 "Doors and windows", SC 3 "Hardware", WG 3 "Terminology for window mechanisms"/TC 162 "Portes et fenêtres", SC 3 "Quincaillerie", GT 3 "Terminologie des dispositifs de fermeture de fenêtre"
ISO/TC 162/SC 3/WG 3
Parent organization: International Organization for Standardization
Contact: Österreichisches Normungsinstitut (ON)
Heinestr. 38
Postfach 130
1021 Wien 2
Austria
Vocabulary standards:
Subject(s): doors; windows
UDC 69. 028
Language(s): En; Fr; Ru

● ***254*** *(Commission/Committee)*
TC 162 "Doors and windows", SC 3 "Hardware", WG 4 "Terminology for door and window fittings"/ TC 162 "Portes et fenêtres", SC 3 "Quincaillerie", GT 4 "Terminologie des accessoires pour portes et fenêtres"
ISO/TC 162/SC 3/WG 4
Parent organization: International Organization for Standardization
Contact: DIN Deutsches Institut für Normung
Burggrafenstraße 4–10
Postfach 1107
1000 Berlin 30, FRG
Vocabulary standards:
Subject(s): doors; windows
UDC 692. 8
Language(s): En; Fr; Ru

● ***255*** *(Commission/Committee)*
TC 163 "Thermal insulation", WG 1 "Terminology"/ TC 163 "Isolation thermique", GT 1 "Terminologie"
ISO/TC 163/WG 1
Parent organization: International Organization for Standardization
Contact: American National Standards Institute (ANSI)
1430 Broadway
New York
N. Y. 10018
USA
Vocabulary standards:
Subject(s): thermal insulation
UDC 662. 998
Language(s): En; Fr; Ru

● ***256*** *(Commission/Committee)*
TC 164 "Mechanical testing of metals", WG 1 "Terminology and symbols"/TC 164 "Essais mé caniques des métaux", GT 1 "Terminologie et symboles"
ISO/TC 164/WG 1
Parent organization: International Organization for Standardization
Contact: American National Standards Institute (ANSI)
1430 Broadway
New York
N. Y. 10018
USA
Vocabulary standards:
Subject(s): mechanical testing (of metals); metals (mechanical testing)
UDC 669; UDC 620. 17
Language(s): En; Fr; Ru

● ***257*** *(Commission/Committee)*
TC 166 "Ceramic ware, glassware and glass ceramic ware in contact with food", WG 1 "Terminology"/TC 166 "Articles en céramique, en verre et en céramique vitreuse en contact avec les denrées alimentaires", GT 1 "Terminologie"
ISO/TC 166/WG 1
Parent organization: International Organization for Standardization
Contact: British Standards Institution (BSI)
2 Park Street
London W1A 2BS
UK
Vocabulary standards:
Subject(s): ceramic ware; glassware; glass ceramic ware
UDC 666; UDC 641
Language(s): En; Fr; Ru

● ***258*** *(Commission/Committee)*
TC 168 "Prosthetics and orthotics", WG 1 "Nomenclature and classification" TC 168 "Prothèses et orthèses", GT 1 "Nomenclature et classification"
ISO/TC 168/WG 1
Parent organization: International Organization for Standardization
Contact: American National Standards Institute (ANSI)
1430 Broadway
New York
N. Y. 10018
USA
Vocabulary standards:
Subject(s): prosthetics; orthotics
UDC 616–089. 23
Language(s): En; Fr; Ru

● ***259*** *(Commission/Committee)*
TC 171 "Micrographics", WG 5 "Vocabulary"/TC 171 "Micrographie", GT 5 "Vocabulaire"
ISO/TC 171/WG 5
Parent organization: International Organization for Standardization

Contact: Association française de normalisation (AFNOR)
Tour Europe
Cedex 7
92080 Paris La Défense
France
Vocabulary standards:
Subject(s): micrographics
UDC 778. 14
Language(s): En; Fr; Ru

● ***260*** *(Commission/Committee)*
TC 173 "Technical systems and aids for disabled or handicapped persons", SC 2 "Classification and terminology"/TC 173 "Assistances et aides techniques pour les invalides ou handicapés", SC 2 "Classification et terminologie"
ISO/TC 173/SC 2
Parent organization: International Organization for Standardization
Contact: Dansk Standardiseringsraad (DS)
Aurehøjvej 12
Postbox 77
DK-2900 Hellerup
Denmark
Vocabulary standards:
Subject(s): handicapped persons (aids for); disabled persons (aids for)
UDC 615. 477
Language(s): En; Fr; Ru

● ***261*** *(Commission/Committee)*
TC 176 "Quality assurance", WG 1 "Quality assurance terminology"/TC 176 "Assurance de la qualité", GT 1 "Terminologie sur l'assurance de la qualité"
ISO/TC 176/WG 1
Parent organization: International Organization for Standardization
Contact: Association française de normalisation (AFNOR)
Tour Europe
Cedex 7
92080 Paris La Défense
France
Vocabulary standards:
Subject(s): quality assurance
UDC 658. 56
Language(s): En; Fr; Ru

● ***262*** *(Commission/Committee)*
TC 180 "Solar energy", WG 1 "Nomenclature"/ TC 180 "Energie solaire", GT 1 "Nomenclature"
ISO/TC 180/WG 1
Parent organization: International Organization for Standardization
Contact: Standards Council of Canada (SCC)
International Standardization Branch
2000 Argentia Road
Suite 2–401
Mississauga, Ontario L5N 1V8, Canada
Vocabulary standards:
Subject(s): solar energy
UDC 662. 997
Language(s): En; Fr; Ru

◐ ***263*** *(Organization)*
International Pharmaceutical Abstracts
IPA
Parent organization: American Society of Hospital Pharmacists
4630 Montgomery Avenue — Phone: (301) 657–3000
Bethesda, Md., 20814
USA
Staff: 12 (does not include records persons, etc.)
Head: Dwight R. Tousignaut, Pharm. D.
Expert(s): Dwight R. Tousignaut, Pharm. D.
Staff categories:
2 terminologist(s) (special field); 1 documentalist(s)/librarian(s); 7 other professional(s) (abstractors/indexers)
Founded: 1964
Activities: No terminological activities as such
Subject(s): pharmacy; pharmaceuticals (technology and use)
UDC 615
Language(s): En
Publications: *International Pharmaceutical Abstracts (IPA)*; *IPA Users Guide*; *IPA Thesaurus of Subject Terms*; *Drug Trade Name Cross Reference List*; IPA 3-Year Cumulative Subject/Author Index (1978, 1979, 1980)
International Pharmaceutical Abstracts, a comprehensive reference network for researching any drug-related subject. The IPA Information System offers an entire spectrum of technical information services
Services: reproduction; translation (all abstracts are in En); referral
Service charges (for reproduction services)
Operation:
computerized
Data base description:
Name: IPA Information System
Data processing:
Hardware: IBM
Access:
on-line (through TOXLINE (National Library of Medicine) and DIALOG (Lockheed)); off-line
Data available:
with charge

(Project(s))
Responsible: International Pharmaceutical Abstracts
Title: Assigning CAS Registry numbers and CODES to IPA
Principal investigator: Dwight R. Tousignaut, Pharm. D.
Collaborators: 8
Publications: Thesaurus; magnetic tapes

● ***264*** *(Organization)*
International Telecommunication Union/Union internationale des télécommunications
ITU/UIT
Place des Nations — Phone: (022) 99. 51. 11
Genève 20, CH-1211 — Telegr.: Burinterna Genève
Switzerland
Staff:
Head: Mr. G. Byrne-Sutton, Chief of the Language Division
Expert(s): Mr. A. Descalzi, terminology co-ordinator
Founded: 1865
Documentation: bibliographic data
Subject(s): telecommunication
UDC 621. 39
Language(s): En; Fr; Es
Vocabularies:
Subject(s): telecommunication
UDC 621. 39
Language(s): En; Fr; Es
Publications: *International Telecommunication Convention*, Final Protocol, Additional Protocols, Resolutions, Recommendations and Opinions (Malaga-Torremolinos 1973); Radio Regulations; Publications of the International Telegraph & Telephone Consultative Committee (CCITT): *Yellow Book*, VIIth Plenary Assembly, Geneva, 1980; Yellow Book list of terms and definitions; Publications of the International Radio Consultative Committee (CCIR): XVth Plenary Assembly, Geneva, 1982;
Provisional Glossary of Telecommunications Terms. Vol. I English into French and Spanish (out of print); Vol. II French into English and Spanish (out of print); Vol. III Spanish into English and French 1979; *Radiocommunication Terms* (GLOS 80/1); *Telecommunication Terms*

(GLOS 80/2); trilingual En, Fr, Es, 1981; *Mobile Services* (GLOS/3) trilingual with definitions En, Fr, Es, 1982; *Terms relating to the administrative and financial terms* Part 1 (GLOS 4) trilingual with definitions in En, Fr, Es, 1982, trilingual with definitions En, Fr, Es, 1982.

● ***265*** *(Organization)*
International Union of Crystallography
I. U. Cr.
5 Abbey Square
Chester, CH1 2HU
UK
Phone: 44 244 (Chester) 42 878
Telex: 667325 COMCABG, attn. Unicrystal
Telegr.: Unicrystal, Chester, England
Staff: 9
Head: Dr. J. N. King, executive secretary; Prof. J. Karle, president
Expert(s): Dr. S. C. Abrahams, Chairman of the IUCr Commission on Journals, and the IUCr Commission on Crystallographic Nomenclature
Founded: 1949
Documentation:
bibliographic data
factographic data
Subject(s): crystallography
UDC 548
Language(s): En (usually)
Publications: *Journal of Applied Crystallography* (first published in 1968) and *Acta Crystallographica* (first published in 1948) by Munksgaard, Copenhagen; Structure Reports, published by Reidel, Dordrecht-series ; *International Tables for X-ray Crystallography* (Vol.I to IV), published by the Kynoch Press, Birmingham;
Bibliographies: *High Temperature X-ray Diffraction Techniques* (1964), *Low Temperature X-ray Diffraction* (1964), *Methods of Obtaining Monochromatic X-rays and Neutrons* (1967), *Diffusion des Rayons X aux Petits Angles* (1970). The titles of books published by the IUCr can be obtained from Dr. King.

● ***266*** *(Commission/Committee)*
Ad Hoc Committee on the Nomenclature of Disordered, Modulated and Polytype Structures
Parent organization: International Union of Crystallography
5 Abbey Square
Chester, CH1 2HU
UK
Phone: 44 244 (Chester) 42878
Telex: 667325 COMCABG, attn. Unicrystal
Telegr.: Unicrystal, Chester, England
Contact: Dr. S. C. Abrahams
Bell Laboratories
Murray Hill
NJ 07974
USA
Committee members: 9
Chairperson: Prof. A. Guinier
Established: 1979

● ***267*** *(Commission/Committee)*
Ad Hoc Committee on the Nomenclature of Symmetry
Parent organization: International Union of Crystallography
5 Abbey Square
Chester, CH1 2HU
UK
Phone: 44 244 (Chester) 42878
Telex: 667325 COMCABG, attn. Unicrystal
Telegr.: Unicrystal, Chester, England
Contact: Dr. S. C. Abrahams
Bell Laboratories
Murray Hill
NJ 07974
USA
Committee members: 8
Chairperson: Prof. P. N. de Wolff
Established: 1980

● ***268*** *(Organization)*
International Union of Physiological Sciences' Board for the Definition of Terms
IUPS
Institute of Arctic Biology,
University of Alaska
Fairbanks, Alaska, 99701
USA
Phone: (907) 479–7648
Staff:
Head: Dr. John Bligh, Chairman
Vocabularies: It is intended to prepare an internationally acceptable glossary of terms for each of the major divisions of physiology. Work is to be carried out by IUPS commissions for the various specialized fields of physiology.
Subject(s): physiology
UDC 612
Language(s): En (and others)

● ***269*** *(Organization)*
International Union of Pure and Applied Chemistry
IUPAC
IUPAC Secretariat,
Bank Court Chambers,
2 - 3 Pound Way
Cowley Centre, Oxford,
OX4 3YF
UK
Phone: Oxford 770125/ 772834
Telex: 83147 attn. IUPAC
Telegr.: IUPAC OXFORD
Staff: 7
Head: Dr. M. Williams, executive secretary
Staff categories:
7 other professional(s)
Founded: 1919
Documentation:
bibliographic data
factographic data
Subject(s): chemistry
UDC 54
Language(s): En (primarily)
Vocabularies: Some 80 projects on chemical terminology underway
Subject(s): chemistry
UDC 54
Language(s): En
Vocabulary standards:
Subject(s): chemistry
UDC 54
Language(s): En
Publications: IUPAC official journal *Pure and Applied Chemistry*. Oxford/N. Y.: Pergamon Press. Recommendations published in the official IUPAC journal and in compendium book-form; Chemistry International, official news magazine of IUPAC, published six times a year.
Services: reference; loan; reproduction; translation (possible)
Service charges
User aids: *Chemistry International* sometimes serves as newsletter

(Project(s))
Responsible: Commission on Nomenclature of Organic Chemistry
Title: Rules of Nomenclature of Organic Chemistry
Project type: commissioned project; in progress
Scope and/or subject matter: Organic chemistry
Publications: IUPAC *Nomenclature of Organic Chemistry*, Sections A, B, C, D, E, F and H, 1979 Edition, Pergamon Press, Oxford 1979; Revision of the *Extended Hantzsch-Widmann System of Nomenclature for Heteromonocycles.*; *Designation of Non-standard Classical Valence Bonding in Organic Nomenclature*; *Revised Nomenclature for Free Radicals, Ions and Radical Ions; Systematic Nomenclature for Delocalized Ions and Radicals; Nomenclature of*

Cyclophanes; Radial Nomenclature; Parent Structures and Derived Radial Substituent Names; Proposals Related to Name Construction

● *270* *(Commission/Committee)*
Commission on Analytical Nomenclature
Parent organization: International Union of Pure and Applied Chemistry (IUPAC)
2 - 3 Pound Way,
Cowley Centre — Phone: 0865/77 01 25
Oxford, OX4 3YF — Telex: 83147 attn. IUPAC
UK — Telegr.: IUPAC OXFORD
Committee members: 16
Chairperson: Prof. G. G. Guilbault (1973 - 1983)
Established: 1959
Vocabularies:
Subject(s): analytical chemistry
UDC 543
Language(s): En

● *271* *(Commission/Committee)*
Commission on Macromolecular Nomenclature
Parent organization: International Union of Pure and Applied Chemistry (IUPAC)
IUPAC Secretariat,
Bank Court Chambers,
2 - 3 Pound Way — Phone: Oxford 770125/
Cowley Centre, Oxford, — 772834
OX4 3YF — Telex: 83147 attn. IUPAC
UK — Telegr.: IUPAC OXFORD
Committee members: 16
Chairperson: Prof. A. D. Jenkins (1975 - 1983)
Established: 1968
Vocabularies:
Subject(s): macromolecular chemistry
UDC 541. 64
Language(s): En

● *272* *(Commission/Committee)*
Commission on Nomenclature of Inorganic Chemistry
Parent organization: International Union of Pure and Applied Chemistry (IUPAC)
IUPAC Secretariat,
Bank Court Chambers,
2 - 3 Pound Way — Phone: Oxford 770125/
Cowley Centre, Oxford, — 772834
OX4 3YF — Telex: 83147 attn. IUPAC
UK — Telegr.: IUPAC OXFORD
Committee members: 17
Chairperson: Prof. Y. Jeannin (1981–1983)
Vice-chairperson: Prof. D. H. Busch (1981–1983)
Established: 1921
Vocabularies:
Subject(s): inorganic chemistry
UDC 546
Language(s): En

● *273* *(Commission/Committee)*
Commission on Nomenclature of Organic Chemistry
CNOC
Parent organization: International Union of Pure and Applied Chemistry (IUPAC)
2 - 3 Pound Way,
Cowley Centre — Phone: 0865/77 01 25
Oxford, OX4 3YF — Telex: 83147 attn. IUPAC
UK — Telegr.: IUPAC OXFORD
Committee members: 17
Chairperson: Dr. G. P. Moss (1981–1983)
Vice-chairperson: Prof. D. Hellwinkel (1981–1983)
Established: 1934
Vocabularies:
Subject(s): organic chemistry
UDC 547
Language(s): En
Publications: *IUPAC Nomenclature of Organic Chemistry*, Sections A, B, C, D, E, F and H, Edition Pergamon Press, Oxford, 1979.

● *274* *(Commission/Committee)*
Commission on Physicochemical Symbols, Terminology and Units
Parent organization: International Union of Pure and Applied Chemistry (IUPAC)
IUPAC Secretariat,
Bank Court Chambers,
2 - 3 Pound Way — Phone: Oxford 770125/
Cowley Centre, Oxford, — 772834
OX4 3YF — Telex: 83147 attn. IUPAC
UK — Telegr.: IUPAC OXFORD
Committee members: 12
Chairperson: Prof. K. Kuchitsu (1981–1983)
Established: 1936
Vocabularies:
Subject(s): physical chemistry
UDC 541. 1
Language(s): En

● *275* *(Commission/Committee)*
Commission on Quantities and Units
Parent organization: International Union of Pure and Applied Chemistry (IUPAC)
2 - 3 Pound Way, Cowley Centre
Oxford, OX4 3YF — Telex: 83147 attn. IUPAC
UK — Telegr.: IUPAC OXFORD
Committee members: 9
Chairperson: Dr. H. P. Lehmann (1980–1983)
Established: 1967
Vocabularies:
Subject(s): analytical chemistry
UDC 543
Language(s): En

● *276* *(Commission/Committee)*
Interdivisional Committee on Nomenclature and Symbols
IDCNS
Parent organization: International Union of Pure and Applied Chemistry (IUPAC)
IUPAC Secretariat,
Bank Court Chambers,
2 - 3 Pound Way — Phone: Oxford 770125/
Cowley Centre, Oxford, — 772834
OX4 3YF — Telex: 83147 attn. IUPAC
UK — Telegr.: IUPAC OXFORD
Committee members: 15
Chairperson: Dr. K. L. Loening (1976 - 1984)
Vice-chairperson: Prof. N. Lozac'h (1975 - 1983)
Established: 1975
Vocabularies:
Subject(s): chemistry
UDC 54
Language(s): En

◐ *277* *(Organization)*
Internationale Gesellschaft für Getreidechemie/International Association for Cereal Chemistry/Association internationale de chimie céréalière
ICC
Schmidgasse 3–7 — Phone: (0222) 77 72 02
Schwechat, A-2320 — Telex: 13 33 16
Austria — Telegr.: INTERCERO-CHEM WIEN
Head: Dipl. Ing. Helmut Glattes
Expert(s): Prof. Dr. R. Schneeweiss
Founded: 1955
Activities: Only marginally concerned with terminology
Documentation: bibliographic data
Subject(s): cereal chemistry; cereal products
UDC 633. 1; UDC 663; UDC 664. 6/. 7
Language(s): En; Fr; De; Ru
Publications: *Standard Methods*; *ICC Congress Reports*; *Synoptic Tables*

(Project(s))
Responsible: ICC Working Group (12) Documentation - Information

Title: Multilingual glossary of cereal chemistry and cereal processing
Principal investigator: Prof. Dr. R. Schneeweiss
Scope and/or subject matter: Cereal chemistry; cereal processing

● ***278*** *(Commission/Committee)*

ICC Working Group (12) Documentation - Information

Parent organization: International Association for Cereal Chemistry
Schmidgasse 3–7
Schwechat, A-2320
Austria
Phone: (0222) 77 72 02
Telex: 13 33 16
Telegr.: INTERCERO-CHEM WIEN
Contact: Prof. Dr. R. Schneeweiss
Direktor des Instituts für Getreideverarbeitung
Arthur Scheunert-Allee 40–41
1505 Bergholz-Rehbrücke
GDR
Committee members: 34
Chairperson: Prof. Dr. R. Schneeweiss
Vice-chairperson: Ing. J. Bartsch
Established: 1956
Vocabularies:
Subject(s): cereal chemistry (and related fields)
UDC 633. 1; UDC 663; UDC 664. 6/. 7
Language(s): En; Fr; De; Ru
Publications: Multilingual glossary of cereal chemistry and cereal processing (new edition underway)

● ***279*** *(Organization)*

Internationales Institut für Rechts- und Verwaltungssprache

[International Institute for Legal and Administrative Terminology; Institut international de terminologie juridique et administrative; Instituto internacional de terminología jurídica y administrativa (Es); Istituto internazionale per la terminologia giuridica ed amministrativa (It)]
Knesebeckstraße 8 - 9
Berlin 12, D-1000
FRG
Phone: 030/867 4307
Telex: 41 183798
Staff: 7
Head: Dr. Alexander Lane
Expert(s): Dr. Alexander Lane
Staff categories:
1 terminologist(s) (special field); 1 linguist(s); 5 other professional(s)
Founded: 1963
Subject(s): law; administration
UDC 34; UDC 35
Language(s): En; Fr; De; It; Es
Vocabularies:
Subject(s): law; administration
UDC 34; UDC 35
Language(s): En; Fr; De; It; Es
Publications: *Europa-Glossar* (25 volumes) der Rechts- und Verwaltungssprache. Verlag Langenscheidt KG.
Band 1. *Büro und Geschäftsgang, Telefon- und Telegrammverkehr.* De/Fr.
Band 2. *Verhandlungssprache.* De/Fr.
Band 3. *Besoldung.* De/Fr. Band 4. *Allgemeines Verwaltungsrecht und -streitverfahren.* De/Fr.
Band 5. *Verwaltungsorganisation - Amtsbezeichnungen - Behörden.* De/Fr.
Band 6. *Kommunalrecht.* De/Fr.
Band 7. *Haushalt.* De/Fr. Band 8. *Beamtenrecht.* De/Fr.
Band 9. *Raumordnung.* De-Fr/Fr-De. In co-ordination with the Council of Europe.
Band 10. *Öffentliches Auftragswesen.* De/Fr.
Band 11. *Jugendfragen.* De/Fr.
Band 12. *Ausgewählte Rechtsbegriffe aus europäischen Verträgen.* Fr/En. In co-ordination with the Council of Europe.
Band 13. *Ausgewählte Begriffe des Niederlassungsrechts.* De/En. In co-ordination with the Council of Europe.
Band 14. *Kommunalrecht.* De/En.
Band 15. *Begriffe des englischen Vertragsrechts.* En/Fr. In co-ordination with the Council of Europe.
Band 16. *Öffentliche Betriebe.* De/Fr.
Band 17. *Raumordnung.* De/It.
Band 18. *Raumordnung.* De/En.
Band 19. *Eherecht.* De/Fr.
Band 20. *Kommunalrecht.* De/It.
Band 21. *Niederlassungsrecht.* De/Fr.
Band 22. *Verwaltungsorganisation.* De/It.
\+ Band 23. *Bildungswesen.* De/Fr.
Band 24. *Büro und Geschäftsgang, Telefon- und Telegrammverkehr.* De/It.
Band 25. *Verhandlungssprache.* De/It.
Forthcoming: Band 26. *Haushalt und Rechnungsprüfung.* De/It. Band 27. *Kraftfahrtversicherung.* De/En. Band 28. *Haushalt und Rechnungsprüfung.* De/Es. Band 29. *Umweltpolitik.* De/En.
Terminological data:
definitions (partly); equivalents

● ***280*** *(Commission/Committee)*

IUPAC-IUB Joint Commission on Biochemical Nomenclature
JCBN

Parent organization: International Union of Pure and Applied Chemistry (IUPAC) and International Union of Biochemistry (IUB)
Bank Court Chambers, 2–3 Pound Way, Cowley Centre
Oxford, OX4 3YF
UK
Contact: Dr. H. B. F. Dixon
Secretary of the Commission
Dept. of Biochemistry
Tennis Court Rd., Cambridge CB2 1QW, U. K.
Committee members: 8
Chairperson: Prof. Dr. rer. nat. Peter Karlson
Established: 1977
Vocabulary standards:
Subject(s): biochemistry; chemistry (of natural products)
UDC 54; UDC 577. 1
Language(s): En
Publications: NC-IUB and JCBN Newsletter; *Biochemical Nomenclature and Related Documents* (1978); *Enzyme Nomenclature* (1979); *Amino-acid Nomenclature,* etc.; without charge

● ***281*** *(Commission/Committee)*

Nomenclature Committee of IUB
NC-IUB

Parent organization: International Union of Biochemistry
P. O. Box 016129
Miami, Florida 33101
Dr. W. J. Whelan, General Secretary of IUB, Biochemistry, USA
Contact: Dr. H. B. F. Dixon
Secretary of the Nomenclature Commission
Dept. of Biochemistry
Tennis Court Rd., Cambridge CB2 1QW, U. K.
Committee members: 8
Established: 1977
Vocabulary standards:
Subject(s): biochemistry (and related fields)
UDC 577. 1
Language(s): En
Publications: NC-IUB and JCBN Newsletter; *Biochemical Nomenclature and Related Documents* (1978); *Enzyme Nomenclature* (1979); *Amino-acid Nomenclature,* etc. (without charge)

● ***282*** *(Organization)*

Organisation internationale de métrologie légale

[International Organization of Legal Metrology]
OIML; IOLM
11, rue Turgot
Paris, F-75009
France
Phone: (1) 878 12 82; 285 27 11
Telex: 660 870 SVP SERV – Code 1103

Staff:
Head: Mr. B. Athané, Director, International Bureau of Legal Metrology (BIML)
Founded: 1955
Activities: Carried out by international working groups named Pilot Secretariats and Reporting Secretariats composed each of several member-countries and international institutions
Publications: Publications are published in the official language of OIML, i. e. in French. Most of the publications are available in English as well. Some of them have been translated into several other languages. *Bulletin de l'Organisation internationale de métrologie légale*, a periodical containing news and technical papers of interest to the national metrology services.
International Vocabulary of Metrology, prepared in cooperation with ISO, IEC and BIPM.
Services: reference; reproduction; translation
No service charges (for Member States)

● ***283*** *(Commission/Committee)*
Pilot Secretariat (SP) **1 "Terminology** (of measurements)"
Parent organization: Organisation Internationale de Métrologie Légale

● ***284*** *(Commission/Committee)*
SP 1–Sr 1 "Vocabulary of legal metrology, fundamental terms"
Parent organization: Organisation Internationale de Métrologie Légale

● ***285*** *(Commission/Committee)*
SP 1–Sr 2 "Vocabularies of various measurement fields (general questions)"
Parent organization: Organisation Internationale de Métrologie Légale

● ***286*** *(Commission/Committee)*
SP 1–Sr 3 "Terminological conformity of OIML documents"
Parent organization: Organisation Internationale de Métrologie Légale

● ***287*** *(Commission/Committee)*
SP 4–Sr 7 "Terminology of length and angle measurements"
Parent organization: Organisation Internationale de Métrologie Légale (in co-operation with ISO)

● ***288*** *(Commission/Committee)*
SP 5–Sr 1 "Terminology of volume measurements of liquids and gases"
Parent organization: Organisation Internationale de Métrologie Légale

● ***289*** *(Commission/Committee)*
SP 7–Sr 1 "Terminology of mass measurements"
Parent organization: Organisation Internationale de Métrologie Légale

● ***290*** *(Commission/Committee)*
SP 9–Sr 9 "Terminology of density measurements"
Parent organization: Organisation Internationale de Métrologie Légale

● ***291*** *(Commission/Committee)*
SP 12–Sr 1 "Terminology of temperature and heat measurements"
Parent organization: Organisation Internationale de Métrologie Légale

● ***292*** *(Commission/Committee)*
SP 13–Sr 7 "Terminology of electrical and magnetic measurements"
Parent organization: Organisation Internationale de Métrologie Légale

● ***293*** *(Commission/Committee)*
SP 19–Sr 6 "Terminology of force and hardness measurements"
Parent organization: Organisation Internationale de Métrologie Légale

● ***294*** *(Commission/Committee)*
SP 27–Sr 1 "Terminology of reference materials"
Parent organization: Organisation Internationale de Métrologie Légale

● ***295*** *(Organization)*
Unesco Social Science Documentation Centre and DARE Data Bank/Centre de documentation en sciences sociales de l'Unesco/Centro de documentación en ciencias sociales de la Unesco (Es)
Unesco/SSDC
7, Place de Fontenoy
Paris, F-75700
France
Phone: National (1) 577–16–10 ext. 59–35 (34, 35); international: + 33 1 577–16–10 ext. 59–35
Telex: 204461 Paris
Telegr.: Unesco Paris
Visiting address: 1, rue Miollis, Paris 15e, or 35, rue François Bonvin, Paris 15e
Staff: 7
Head: Mr. Santos Cuyugan, Director
Expert(s): Ester M. Ronquillo, Programme Specialist
Staff categories:
1 documentalist(s)/librarian(s); 1 information specialist(s)
Founded: 1952
Documentation:
bibliographic data
factographic data
Subject(s): social sciences
UDC 3
Language(s): En; Fr; Es; Pt; Ar; Ru
Vocabularies:
Subject(s): social sciences
UDC 3
Language(s): En
Publications: Gould, J.; Kolb, W. L. (eds). *A Dictionary of Social Sciences.* London: Tavistock, 1964, 761 p. Unesco.
Diccionario de ciencias sociales V. I. Unesco; s. e., 1975. Compiled by the Grupo Español de Trabajo composed by the Comisón Nacional de Cooperación con la Unesco - Instituto de Estudios Políticos (Madrid). In preparation: Portuguese and Arabic dictionaries, and an *Integrated Thesaurus of Social Sciences;*
VIET, J. *Thesaurus for Information Processing in Sociology/Thésaurus pour le traitement de l'information en sociologie.* (Prepared under the auspices of the International Committee for Social Sciences Information and Documentation and with the financial support of Unesco). Paris: Mouton, 1972.
World List of Social Science Periodicals, 5th ed., 1980; *World Directory of Social Science Institutions*, 2nd ed., 1980
Services: reference; referral
No service charges
Operation:
manual
computer-assisted
Data base description: 9,000 records on social science institutions, specialists, research, periodicals, data services, etc.
Name: DARE Data Bank
Size: 9, 000 records
Language(s): En
Classification: UDC; DARE Descriptor List; Unesco Thesaurus; OECD Macrothesaurus
Terminological data:
definitions; explanations; equivalents
Data processing:
Hardware: IBM; 3031 Central processing unit; with an

internal memory of 2048 Kbytes; External storage comprising 5 disc drives (10 spandles) allowing direct access to 3, 170 Kbytes, 6 tape units handling data densities of 800/1, 600/6, 250 bytes per inch; two line printers (2, 000 and 1, 200 lines per minute) and a teleprocessing system including line controllers modules and terminals (CRT, printers and typewriters); Two terminals "Memorex 1377" and one "Delta 5500" visual display plus a GE Terminet 1800 printer. On order a third Memorex terminal and an Olivetti electronic text processor with screen ET 221.
Software: operating system: ISIS/Unesco; operation mode: real time; batch; programming language(s): PL1; Assembler; data base management system/retrieval documentation system: DARE/ISIS/Unesco
Data available:
on exchange basis; free of charge

● *296* *(Organization)*
Unesco. Clearing House and Research Centre for Cultural Development/Unesco. Centre de documentation d'information et de recherche pour le développement culturel/Unesco. Centro de documentación, información y investigaciones sobre el desarrollo cultural
7, Place de Fontenoy
Paris, F-75700
France
Phone: 577–16–10
Telex: 204461 Paris
Telegr.: UNESCO PARIS
Staff: 4 (part time)
Head: Ms. G. de La Rochefoucauld (head)
Staff categories:
1 documentalist(s)/librarian(s)
Founded: 1971
Documentation: bibliographic data
Subject(s): cultural development
UDC 008
Language(s): En; Fr
Vocabularies:
Subject(s): cultural development
UDC 008
Language(s): En; Fr; Es
Publications: Viet, J. *International thesaurus of cultural development/Thésaurus international du développement culturel.* Paris: Unesco, 1980, 498 p.
Services: reference; referral
Operation:
manual
computer-assisted (Part of Unesco Computerized Documentation System)

● *297* *(Organization)*
Unesco. Communication Documentation Centre/ Unesco. Centre de documentation de la communication
CC/FCP/DCD
7, Place de Fontenoy
Paris, F-75700
France
Phone: 577–16–10
Telex: 204461 Paris
Telegr.: UNESCO PARIS
Staff: 3
Expert(s): Ms. Janine Gimon-Bastoul
Staff categories:
1 documentalist(s)/librarian(s); 2 other professional (s)
Founded: 1951
Documentation: bibliographic data
Subject(s): mass communication
UDC 659. 3
Language(s): En; Fr; Es
Vocabularies:
Subject(s): mass communication
UDC 659. 3
Language(s): En; Fr; Es
Publications: VIET, J. *Thesaurus Mass Communication/ Communication de masse/Comunicación colectiva.* Paris: Unesco, 1975, 148 p. Alphabetical list of descriptors in En, Fr, Es in course of revision
Annual bibliographies: *List of Documents and Publications in the field of Mass Communication*
Services: reference
No service charges
Operation:
computerized

● *298* *(Organization)*
Unesco. Translation and Records Division; Terminology and Documentation Section/Unesco. Division des Traductions et comptes rendus; Section de terminologie et documentation
COLT
Parent organization: UNESCO
7 Place de Fontenoy
Paris, F-75700
France
Phone: 577–16–10 ext. 30–17
Telex: 270602 Paris
Telegr.: UNESCO PARIS
Staff: 9
Head: Mr. Luis G. Marqués, Director
Expert(s): Dr. C. M. Gatfield
Staff categories:
1 terminologist(s) (theory); 8 other professional(s) (A lexicographer, chief of section; 7 reference and clerical staff) plus occasional assistance from translators
Founded: 1975 (reestablished)
Documentation:
bibliographic data
factographic data
Subject(s): education; culture; communications; science; social sciences
UDC 0/6
Language(s): En; Fr; Es; Ru; Ar; Ch
Vocabularies:
Subject(s): education; culture; communications; science; social sciences
UDC 0/6
Language(s): En; Fr; Es; Rr; Ar; Ch
Publications: *Termes de la Caisse commune des pensions du personnel des Nations Unies;* in En, Fr, Es (PMTL.3); *Liste de termes administratifs et financiers;* in En, Fr, Es, Ru (COLT/PMTL/3), updating in progress; *Glossary of seismic terms in 4 languages* compiled by the Interpretation Division (COL-T/TERM/1); Liste de termes sur les *Politiques de la communication*; in En, Fr, Es, Ru (COL-T/TERM/3); Liste de termes sur les *Politiques culturelles*: in En, Fr, Ru (COL-T/TERM/4); List of NGOs in categories A, B and C in 5 languages (COL-T/TERM/5) - updated version in 6 languages to be published shortly; Updated version of French, Spanish and Russian indexes of the above list of NGOs to be published shortly; List of Intergovernmental Organizations with which Unesco has established relations in five languages (COL-T/TERM/6) - updated version in 6 languages to be published shortly; Detailed structure of Unesco's Secretariats and Units away from Headquarters in 6 languages (COL-T/TERM/7); Liste multilingue provisoire de termes d'informatique en 5 langues (COL-T/TERM/9); Titres, sigles et abréviations se rapportant aux activités de l'Unesco en 6 langues (COL-T/TERM/10);
International glossary of hydrology published in 4 languages jointly by Unesco and WMO in 1974
Services: reference; loan; other: terminological answering service
No service charges
Operation:
computer-assisted
Data base description:
Language(s): En; Fr; Es; Ru
Classification: UDC
Data processing:
Hardware: IBM; 3031; 2 Mb; (Disc 3350) 317, 500 Mb; 6; 2 printers, 1 card reader; Terminal Hewlett-Packard with Roman/Cyrillic characters; connected to Unesco computer IBM 3031; Printer: wheel printer/terminal DIABLO 1640 KSR with Roman/Cyrillic characters ordered; delivery expected

(Project(s))
Title: Glossary of education terms

Project type: institutional (internal); in planning
Principal investigator: C. M. Gatfield
Scope and/or subject matter: Education

Title: Glossary of environment terms
Project type: institutional (internal); in planning
Principal investigator: C. M. Gatfield
Scope and/or subject matter: Environment

● ***299*** *(Organization)*
Union internationale de l'industrie du gaz
[International Gas Union; Internationale Gasunion (De); Meždunarodnyj gazovyj sojuz (Ru)]
U. I. I. G. ; I. G. U.; IGU (De)
Parent organization: National gas associations
62, rue de Courcelles Phone: (1) 766 03 51
Paris, F-75008
France
Staff: 15 members on Dictionary Committee
Head: M. Goudal, secrétaire général
Expert(s): Dipl. -Dolm. H. -J. Stellbrink
Staff categories:
2 translator(s); 13 other professional(s)
Founded: 1930
Documentation:
bibliographic data
factographic data
Subject(s): gas engineering
UDC 622. 279; UDC 66. 07
Language(s): En; Fr; De; Ru
Vocabularies:
Subject(s): gas engineering
UDC 622. 279; UDC 66. 07
Language(s): En; Fr; De; Ru
Publications: Committee G reports to World Gas Conference
Dictionary of the Gas Industry in seven languages. Amsterdam: Elsevier Publishing Company. First edition 1961, reprint 1973, reprint 1979; Supplement to the Dictionary of the Gas Industry in seven languages. Amsterdam: Elsevier Scientific Publishing Company first edition 1973; *International Dictionary of the Gas Industry*, edited by Kammer der Technik - GDR, for the International Gas Union, 1981; new *Dictionary* to be published in 1982
Operation:
manual (to be transferred to new data bank in new computer (final selection not yet made))
Terminological data:
definitions; explanations; classified layout; context; equivalents

(Project(s))
Responsible: Sub-Committee G IV
Title: Updating of Dictionary of the Gas Industry (published by Elsevier in 1961)
Project type: commissioned project (International Gas Union); in progress
Principal investigator: Dipl. -Dolm. H. -J. Stellbrink
Collaborators: 15
Scope and/or subject matter: Gas engineering dictionary
Method of work: Input on basis of standardized terms and publications, translations into other languages on same basis
Source material: Belgian, British, French, ISO, U. S. and West German Standards, monolingual dictionaries and handbooks on gas and oil engineering; IGU Reports and Papers
Publications: *Dictionary of the Gas Industry in seven languages.* Amsterdam: Elsevier Scientific Publishing Company. First edition 1961, reprint 1973; reprint 1979; Supplement to the Dictionary of the Gas Industry in seven languages. Amsterdam: Elsevier Scientific Publishing Company, 1973; *International Dictionary of the Gas Industry*, edited by Kammer der Technik - GDR for the International Gas Union, 1981.

● ***300*** *(Commission/Committee)*
Sous-Commission G IV "Dictionnaire"
[Sub-Committee G IV "Dictionary"]
U. I. I. G. G IV; I. G. U. G IV
Parent organization: Union internationale de l'industrie du gaz
62, rue de Courcelles Phone: (1) 766 03 51
Paris, F-75008
France
Contact: Dipl. -Dolm. H. -J. Stellbrink
c/o Ruhrgas AG
Huttropstraße 60
4300 Essen 1, FRG
Committee members: 15
Chairperson: Dipl. -Dolm. H. -J. Stellbrink
Vice-chairperson: Dipl. -Dolm. B. Flamin, secretary
Established: 1950
Vocabularies:
Subject(s): gas engineering
UDC 622. 279; UDC 66. 07
Language(s): En; Fr; Ru; De
Publications: *Dictionary of the Gas Industry in seven languages*

● ***301*** *(Organization)*
Union internationale des chemins de fer
[International Railway Union; Internationaler Eisenbahnverband (De)]
UIC
14 rue Jean Rey Phone: 273. 01. 20
Paris, F-75015 Telex: 270835 Unionfer
France
Staff: 140
Head: M. V. Canyn, Directeur de Bureau de Documentation
Expert(s): Ms. Harvey
Staff categories:
1 terminologist(s) (special field); 2 terminologist(s) (theory); 20 translator(s)
Founded: 1922
Documentation:
bibliographic data
factographic data
Subject(s): railway
UDC 629; UDC 656
Language(s): Fr; En; De
Vocabularies:
Subject(s): railway
UDC 629; UDC 656
Language(s): Fr; En; De; It; Es; Nl
Publications: *General dictionary of railway terms.* In Nl, En, Fr, De, It, Es. 3rd ed., 75 DM. This dictionary is sold by Malsch and Vogel GmbH, Stuttgarter Str. 57c, Karlsruhe, FRG; *Glossary of terms used for mechanical accountancy and electronic data processing machines.* In Fr, with translations in En, De, It, Es. Published provisionally in 1959. 12 FF; *Glossary of signalling terms.* This glossary was compiled in 1972 in En, Fr, De and It and gives a list and translations of signalling terms used on the Railways together with the definition of each term. Cardboard cover with split-ring spring binder enabling new pages to be added and out-of-date pages to be replaced. 45 FF;
Glossary of terms used in connection with supplies. This glossary gives a list of the main terms used by the Railways in connection with supplies in Fr, En, De, It, Pt, Es. Published in 1964. 12 FF; *Glossary of terms used in connection with data transmission.* This glossary contains the terms and definitions used in telecommunications for data transmission in: Cs, En, Fr, De, Hu, It, Pl, Ro, Ru, Sv. Published in 1967. 30 FF; *Glossary of terms used by Permanent Way Departments.* This glossary was compiled in 1975 in En, Fr, De, and It and gives the translations of permanent way terms used on the Railways. 75 FF;
Railway dictionaries. Catalogue according to subject. 1975 ed. available in En, Fr and De. 51 p., 35 FF; *Railway Thesaurus.* Standard language for the use of documentation services. 1979 ed. available in Fr, En and De. 170 p. Price for one edition: 70 FF. Price for three editions

together: 200 FF; *Standard goods nomenclature* (NCM). 1976. Available as
- a list of 4-digit positions -
an analytical list and
- an alphabetical list in En, Fr and De.

(Project(s))
Responsible: Groupe d'automatisation documentaire (GAD)
Title: Introduction d'un système international informatisé de documentation ferroviaire [Introduction of an international computerised railway documentation system]
Project type: commissioned project; in progress
Principal investigator: M. Korving
Collaborators: 8

● ***302*** *(Organization)*
Union postale universelle
[Universal Postal Union]
UPU; UPU
Weltpoststrasse
Case postale — Phone: (+4131) 43 22 11
Berne 15, CH-3000 — Telex: 32 842 UPU CH
Switzerland — Telegr.: UPU Berne
Visiting address: Weltpoststrasse 4
Staff: 135
Head: Mr. Mohamed Ibrahim Sobhi, Director-General of the International Bureau of the UPU
Expert(s): Mr. Emmanuel Nlend
Founded: 1874
Documentation: bibliographic data
Subject(s): postal services; postal law
UDC 656. 801. 313; UDC 351. 816. 1
Language(s): Fr; Ar; Ch; En; De; Ru; Es
Vocabularies:
Subject(s): postal services (international); postal law (international)
UDC 656. 8; UDC 351. 816. 1
Language(s): Fr; Ar; Ch; En; De; Ru; Es
Publications: *Vocabulaire polyglotte du service postal international* [Multilingual vocabulary of the internationl postal services]. In Fr, Ar, Ch, En, De, Ru, Es. This vocabulary contains a selection of terms extracted from the *Actes* and other UPU publications, terms and expressions used in the international postal service as well as terms used in the collection of UPU postal studies. It serves postal employees as well as translators and interpreters.
Operation:
manual
Terminological data:
definitions; explanations; standardized terms; equivalents
Data available:
with charge

● ***303*** *(Commission/Committee)*
Groupe de travail A 19 du Conseil consultatif des études postales (CCEP)
[Working Group A 19 of the Consultative Council for Postal Studies (CCPS)]
GT A 19 du CCEP; WG A 19 of the CCPS
Parent organization: Union postale universelle
Weltpoststrasse
Case postale — Phone: (+ 4131) 43 22 11
Berne 15, CH-3000 — Telex: 32 842 UPU CH
Switzerland — Telegr.: UPU Berne
Visiting address: Berne 15, Weltpoststrasse 4
Contact: Mr. Emmanuel Nlend
Committee members: 3 to 9 countries
Chairperson: (Postal Administration of United Kingdom)
Established: 1952
Vocabularies:
Subject(s): postal services (international)
UDC 656. 8
Language(s): Fr; Ar; Ch; En; De; Ru; Es
Publications: *Vocabulaire polyglotte du service postal international* [Multilingual vocabulary of the international postal service]

● ***304*** *(Organization)*
United Nations Educational, Scientific and Cultural Organization Division of the General Information Programme/Organisation des Nations Unies pour l'éducation, la science et la culture, Division du programme général d'information/Organización de las Naciones Unidas para la educación, la ciencia y la cultura, División del programa general de información
UNESCO
7, Place Fontenoy — Phone: 577-16-10
Paris, F-75700 — Telex: 204461 PARIS
France — Telegr.: UNESCO PARIS
Staff: 42
Head: Mr. Jacques Tocatlian
Expert(s): Dr. Wolfgang Löhner
Staff categories:
2 information specialist(s)
Founded: 1946
Documentation:
bibliographic data
factographic data
Subject(s): information; documentation; libraries; archives
UDC 02; UDC 007
Language(s): En; Fr
Publications: Wüster, E. *Bibliography of monoliqual scientific and technical glossaries. Volume I: National Standards/Bibliographie de vocabulaires scientifiques et techniques monolingues. Volume I : Normes nationales.* Paris: Unesco, 1955, 219 p., 150x240 mm; Wüster, E. *Bibliography of monolingual scientific and technical glossaries. Volume II: Miscellaneous sources/Bibliographie de vocabulaires scientifiques et techniques monolingues. Volume II: Sources diverses. Paris: Unesco, 1959, 146 p., 150x240 mm;* Unesco. *Bibliography of interlingual scientific and technical glossaries. 5th. ed./Bibliographie de dictionnaires scientifiques et techniques multilingues. 5éd./Bibliografia de diccionarios científicos y técnicos plurilingues. 5e ed.* Paris: Unesco, 1969, 250 p., 150x240 mm; Felber H. *Terminological Compatibility: A Basis UNISIST Requirement.* Paris: Unesco, 1972, 20 p. (UNISIST/I/SI/72/1.5) ; *Guidelines for the establishment and development of monolingual thesauri.* Paris: Unesco, 1973, 37 p., A 4. Available in En, Fr, Es. (SC/WS/555); Austin, D.; Dale, P. *Guidelines for the establishment and development of monolingual thesauri.* 2nd revised edition. Paris: Unesco, 1981, 64 p., A4. (PGI-81/WS/15); Liebesny, F. A *State-of-Art Survey on Automatic Indexing.* Paris: Unesco, 1974, 42 p. (COM-74/WS/18); *Indexing Principles.* Paris: Unesco, 1975, 10 p. (SC-75/WS/58); Wersig, G.; Neveling, V. (compilers). *Terminology of documentation/Terminologie de la documentation/ Terminologie der Dokumentation/ Terminologija v oblasti dokumentacii/Terminologia de la documentación:* A selection of 1.200 basic terms published in En, Fr, De, Ru and Es. Paris: Unesco, 1976, 274 p. ISBN 0-85157-140-9;
Guidelines for the establishment and development of multilingual thesauri. Paris: Unesco, 1976, 49 p., A 4 (SC/76/WS/102); *Guidelines for the establishment and development of multilingual thesauri.* Revised text. Paris: Unesco, 1980, 85 p., - A 4 (PGI/80/WS/12; published in En and Fr); *Broad System of Ordering: Schedule and Index.* - 3rd revision. Paris: Unesco; The Hague: International Federation for Documentation, 1978, 184 p.; Felber, H. *Study on the development of a network for terminology information and documentation.* Paris: Unesco, 1978, 45 p, A4.
User aids: Newsletter; Unesco Journal of Information Science, Librarianship and Archives Administration

● ***305*** *(Organization)*
United Nations Environment Programme/Programme des Nations Unies pour l'environnement/Programma organizacii ob"edinennyh nacij po okružajuščej

srede/Programa de las Naciones Unidas para el medio ambiente
UNEP/PNUE/JUNEP/PNUMA

P. O. Box 30552	Phone: 333930
Nairobi	Telex: 22068
Kenya	Telegr.: UNITERRA, Nairobi

Visiting address: Gigiri, Nairobi, Kenya
Staff: 50 (in Conference, Language and Documents Section)
Head: Mr. Mostafa K. Tolba, Executive Director
Expert(s): Mr. Guido Gómez de Silva, Chief, Conference, Language and Documents Section
Staff categories:
11 translator(s); 17 other professional(s)
Founded: 1972
Documentation: bibliographic data
Subject(s): environment
UDC 574
Language(s): Ch; En; Fr, Ru; Es; Ar
Vocabularies:
Subject(s): environment
UDC 574
Language(s): Ch; En; Fr; Ru; Es; Ar
Vocabulary standards:
Subject(s): environment
UDC 574
Language(s): Ch; En; Fr; Ru; Es; Ar
Publications: UNEP/CONF/TERM. 1 *Organizational elements and functional titles in the secretariat of the United Nations Environment Programme* (Ch, En, Fr, Ru, Es), Nairobi, 1980. (ISBN 92-807-0016-2); UNEP/CONF/TERM. 3/Rev. 1 *Names of countries, territories and other areas* (En, Fr, Es), Nairobi, 1981. (ISBN 92-807-0038-3); UNEP/CONF/TERM. 4 *Environmental terminology* (Ch, En, Fr, Ru, Es; Ar), Nairobi, 1982. (ISNB 92-807-0018-9)
Services: reference; translation; other: Documents reproduction
No service charges
Operation:
manual
Data base description:
Name: Terminology cards and glossaries
Size: 1500 cards
Language(s): Ch; En; Fr; Ru; Es
Classification: Alphabetical by term
Terminological data:
equivalents
Access:
via telephone
Data available:
free of charge

● ***306*** *(Organization)*
United Nations Office at Geneva, Terminology and Technical Documentation Section/Office des Nations Unies à Genève, Section de terminologie et de documentation technique
UNOG/ONUG
Parent organization: United Nations Organization, New York

Palais des Nations	Phone: (022) 34 60 11 and 31 02 11, extension 5296
Genève 10, CH-1211	Telex: 289696
Switzerland	Telegr.: UNATIONS, GENÈVE

Visiting address: Palais des Nations, Door 41, Room E. 5102
Staff: 8
Head: Ms. N. Meeus, Chief
Expert(s): Ms. N. Meeus
Staff categories:
5 terminologist(s) (theory); 1 documentalist(s)/librarian(s); 2 other professional(s)
Founded: 1969
Documentation: factographic data
Subject(s): social sciences; natural science; applied sciences
UDC 3/6
Language(s): Ch; En; Fr; Ru; Es; (occasionally Ar)
Vocabularies:
Subject(s): social sciences; natural science; applied sciences
UDC 3/6
Language(s): Ch; En; Fr; Ru; Es (various combinations thereof)
Publications: Terminology bulletins on: *Chemical and bacteriological (biological) weapons*, 1970 31 p., (En, Fr, Ru, Es); *Air and water pollution*, 1970, 37 p., (En, Fr); *Building formalities and procedures*, 1973, 32 p., (En, Fr) ; *International Sugar Agreement*, 1973, 24 p., (En, Fr, Ru, Es); *Détection des explosions nucléaires souterraines*, 1974, 41 p., (En, Fr);
Navigation intérieure, 1974, 114 p., (En, Fr, Ru); *Multimodal transport terminology*, 1975, 105 p., (En, Fr, Ru, Es); *Transfer of technology*, 1976, 69 p., (En, Fr, Ru, Es); *International trade documents*, 1976, 73 p., (En, Fr, Ru, Es); *Cartographie*, 1977, 68 p., (En, Fr); *Human rights terminology*, 1977, 56 p., (En, Fr, Ru, Es); *Jute*, 1978, 37 p., (En, Fr, Ru, Es), plus corr. 1; *Copper industry and trade*, 1979, 65 p., (En, Fr, Ru, Es); *Tea*, 1979, 38 p., (En, Fr, Ru, Es);
Terminology of the natural gas industry, 1980, 67 p., (En, Fr, Es); publications are also available on microfiches; *Glossary on food and agriculture*, 1974, 39 p., (Ch, En); *Terminology relating to armed conflicts*, 1975, 30 p., (Ch, En) *Marine insurance terminology*, 1977, 64 p., (En, Fr, Ru, Es);
Catalogue de la division linguistique, 3ème édition, 1979, 422 p., (En, Fr); *Supplément 2 au catalogue de la division linguistique*, 1980
Services: reference; loan
No service charges (within the organization); outside users are charged for publications unless exchange basis possible
User aids: Internal information circulars
Operation:
computer-assisted
Data processing:
Hardware: IBM; 370/158; 370/3033; 4Mb; 12Mb; 9500Mb; 10 tape drives; 11 RJE stations; The Terminology Section has one of the 200-odd terminals that are linked to the International Computing Centre; it is a Diablo 1620 printer terminal used for input, storage and printing of terminology lists, including Russian, through Wylbur software package
Software: (in International Computing Centre); operating system: OS/MVS; operation mode: batch ; time-sharing; programming language(s): Assembler; PL1; Fortran; Cobol; RPG; Algol; data base management system/retrieval documentation system: Adabas; Inquire; Isis

(Project(s))
Title: Commodity markets
Project type: institutional; in progress
Principal investigator: Mr. S. Silva
Collaborators: 3
Scope and/or subject matter: English-French-Spanish list of terms relating to commodity markets (wheat, cotton, sugar, etc.)
Method of work: Extraction of terminology from books and articles on the subject in each of the languages in question; use of dictionaries, especially monolingual
Source material: Books, journals, documents, dictionaries and experts plus revision of certain existing terminology lists

Title: Transfer of technology
Project type: institutional; in progress
Principal investigator: Mr. S. Lin
Collaborators: 1
Scope and/or subject matter: English-Chinese list of terms relating to the international transfer of technology
Method of work: Extraction of terms from English-language documents and from official Chinese translations done in the People's Republic of China
Source material: United Nations documents

● *307* *(Organization)*

United Nations/Organisation des Nations Unies
[Organización de las Naciones Unidas (Es); Organizacija ob"edinennyh nacii (Ru)]
UN/ONU; ONU (Es); OON Ru)

Grand Central Post Office
Box 20
New York, N. Y., 10017
USA
Phone: (212)754-6597
Telegr.: UNATIONS New York

Staff: 39
Head: Mme. F. Cestac, Director, Translation Division
Expert(s): Dr. David Arnold, Chief, Documentation, Reference, and Terminology Section
Staff categories:
5 terminologist(s) (theory); 3 documentalist(s)/librarian(s); 1 information specialist(s); 30 other professional(s) (1 chief; 29 research assistants)
Founded: 1945
Documentation:
bibliographic data
factographic data
Subject(s): all fields
UDC 0/9
Language(s): Ar; Ch; En; Fr; Ru; Es
Vocabularies:
Subject(s): all fields
UDC 0/9
Language(s): Ar; Ch; En; Fr; Ru; Es
Vocabulary standards:
Subject(s): all fields
UDC 0/9
Language(s): Ar; Ch; En; Fr; Ru; Es
Theory: En; Fr

Publications: Terminology issuances (to wit, translation and editorial aids in the form of multilingual term listings and title indices in various language combinations)
Multilingual and monolingual word lists and indices of titles

Services: reference; loan; translation (but not by the Documentation, Reference and Terminology Section); referral
Service charges: Sales Publications for general public
No charge internal UN

User aids: Terminology Notes

Operation:
manual
computer-assisted

Data base description:
Name: GRIPHOS/CAIP (UNDIS)
Size: 40. 000 records in En/Fr/Ru/Es
Language(s): En
Classification: UDC at present; adoption of a UN thesaurus is under review; alphabetical (nomenclature, agreements, abbreviations, terms); systematic (by topics)

Terminological data:
definitions (optional); explanations (optional); classified layout (optional); context (optional); standardized terms (optional); equivalents (mandatory (this is the purpose of the system)); other data elements (optional)

Data processing:
Hardware: IBM; IV Phase; WANG; IBM 3033 N; WANG OIS140; 8. 4 million bytes (IBM); 15. 200 million bytes (IBM); magnetic tape; high speed printer; printer terminal (used solely as an output device)
Software: operating system: Key to disc to tape (IBM); Key to disc to floppy disk (WANG); operation mode: batch; programming language(s): PL I; data base management system/retrieval documentation system: Shared with other Sections

Access:
via telephone/via telex
off-line

Data available: Through selected publications
on exchange basis

(Project(s))

Title: Abbreviations and Acronyms
Project type: institutional; in progress
Principal investigator: Mme. Cestac (Director, Translation Division)
Collaborators: 9
Scope and/or subject matter: 1. UN organizations and related organizations; 2. International organizations (does not include Non-Governmental Organizations)
Source material: Through official channels

Title: Administrative and Budgetary Terms
Project type: institutional; in progress
Principal investigator: Mme. Cestac (Director, Translation Division)
Collaborators: 9
Scope and/or subject matter: Terms related to administrative and budgetary matters
Source material: 1. UN material (documents, reports, conference papers); 2. Glossaries, terminology bulletins, dictionaries; 3. Technical articles from newspapers, periodicals, books

Title: Disarmament
Project type: institutional; in progress
Principal investigator: Mme. Cestac (Director, Translation Division)
Collaborators: 9
Scope and/or subject matter: Series of UN Conferences on disarmament
Source material: 1. UN material (documents, reports, conference papers); 2. Glossaries, terminology bulletins, dictionaries; 3. Technical articles from newspapers, periodicals, books

Title: European Organizations
Project type: institutional; in progress
Principal investigator: Mme. Cestac (Director, Translation Division)
Collaborators: 9
Scope and/or subject matter: Titles relating to European Organizations
Source material: 1. Material obtained through official channels; 2. Official papers and reports

Title: Law of the Sea
Project type: institutional; in progress
Principal investigator: Mme. Cestac (Director, Translation Division)
Collaborators: 9
Scope and/or subject matter: In connection with UN Conferences on this topic
Source material: All conference related material

Title: List of currencies
Project type: institutional; in progress
Principal investigator: Mme. Cestac (Director, Translation Division)
Collaborators: 9
Scope and/or subject matter: List of official basic units of currencies together with appropriate abbreviations of State Members of UN, members of specialized agencies or parties to the Statute of the International Court of Justice
Source material: Primarily through official channels

Title: Names of Countries and Adjectives of Nationality
Project type: institutional; in progress
Principal investigator: Mme. Cestac (Director, Translation Division)
Collaborators: 4
Scope and/or subject matter: List of States Members of the United Nations
Method of work: Compilation of trilingual lists (En, Fr, Es) of names of countries (full names), adjectives of nationality

inclusive of gender and other useful grammatical indications
Source material: Through offical channels

Title: New and Renewable Sources of Energy (2nd edition)
Project type: institutional; in progress
Principal investigator: Mme. Cestac (Director, Translation Division)
Collaborators: 9
Scope and/or subject matter: As discussed in UN for New and Renewable Sources of Energy
Source material: 1. UN material (documents, reports, conference papers); 2. Glossaries, terminology bulletins, dictionaries; 3. Technical articles from newspapers, periodicals, books

Title: Non-Governmental Organizations
Project type: institutional; in progress
Principal investigator: Mme. Cestac (Director, Translation Division)
Collaborators: 9
Scope and/or subject matter: Index of titles and acronyms of Non-Governmental Organizations
Source material: Through official channels

Title: Peaceful Uses of Outer Space
Project type: institutional; in progress
Principal investigator: Mme. Cestac (Director, Translation Division)
Collaborators: 9
Scope and/or subject matter: United Nations Conference in 1982 on the Peaceful Uses of Outer Space
Source material: 1. UN material (documents, reports, conference papers); 2. Glossaries, terminology bulletins, dictionaries; 3. Technical articles from newspapers, periodicals, books

Title: United Nations Functions and Titles
Project type: institutional; in progress
Principal investigator: Mme. Cestac (Director, Translation Division)
Collaborators: 9
Scope and/or subject matter: Descriptive nomenclature associated with functions and titles in the UN system of organizations
Source material: Through official channels

Title: United Nations Structure
Project type: institutional; in progress
Principal investigator: Mme. Cestac (Director, Translation Division)
Collaborators: 9
Scope and/or subject matter: Official Nomenclature associated with departmental titles in the United Nations
Source material: Through official channels

○ ***308*** *(Organization)*

World Health Organization. Technical Terminology Service/Organisation mondiale de la santé. Service de terminologie technique
WHO/OMS

Avenue Appia
Genève 27, CH-1211
Switzerland
Phone: (022) 34 60 61
Telex: 27821
Telegr.: Unisanté Genève

Staff:
Head: Dr. H. Mahler (Director General of WHO)
Expert(s): Mr. D. Armstrong Lowe (chief, Technical Terminology Service)
Documentation:
bibliographic data
factographic data
Subject(s): public health; physical sciences; biological sciences; medical technology
UDC 61; UDC 614; UDC 57
Language(s): En; Fr; other languages as necessary and as resources permit
Vocabularies:
Subject(s): public health; physical sciences; biological sciences; medical technology
UDC 61; UDC 614; UDC 57
Language(s): En; Fr; other languages as necessary and as resources permit
Publications: WHO has published a large number of recommendations on terminology some of them as books and some of them as articles in various journals. A bibliography of terminology recommendations of interest to public health which includes all WHO publications has been issued as a document in 1976.
A guide to international recommendations on names and symbols for quantities and on units of measurement. Geneva, 1975 (Progress in standardization, no.2); *Guide pour l'utilisation du système international d'unités (SI)*, Genève, 1975; Terminology circulars.
Services: WHO provides a number of services

● ***309*** *(Organization)*

World Meteorological Organization/Organisation météorologique mondiale
WMO/OMM

41, avenue Giuseppe-Motta
P. O. Box 175
Genève, CH-1211
Switzerland
Phone: (011) 34 64 00
Telex: 23260
Telegr.: METEOMOND Genève

Staff: 11
Head: Prof. A. C. Wiin-Nielsen, Secretary-General
Expert(s): Mr. Malcolm Rigby (Chairman of the Working Group on Bibliographic Problems of the Commission for Atmospheric Sciences (CAS)), 5816 22nd St. N., Arlington, Va. 22205, USA
Staff categories:
1 translator(s) (part time)
Founded: 1951 (WMO is the successor to the IMO 1873–1950)
Documentation:
bibliographic data
factographic data
Subject(s): meteorology; climatology; hydrology; oceanography; glaciology; aviation
UDC 551; UDC 556
Language(s): En; Fr; Ru; Es
Vocabularies:
Subject(s): meteorology; climatology; hydrology; oceanography; glaciology, aviation
UDC 551; UDC 556
Language(s): En; Fr; Ru; Es
Publications: *International Meteorological Vocabulary*, 1966 (WMO No.182), 276 p.; 2nd ed. to be published in 1982/83 from E. D. P. base; *International Glossary of Hydrology*, 1st ed., 1974 (WMO No.385), XX 394 p; 1588 terms in En, Fr, Ru, Es, with definitions in these languages are included. They are distributed over the different fields of hydrology concerned with surface water, ground-water, hydrometeorology, solid moisture and related fields such as river hydraulics and water resources management. It also includes *UDC Scheme for General Hydrology* (Class 556) in En and Ru (with explanation for use) as well as tables of recommended symbols and units for hydrological purposes;
Technical Regulations, Volumes I and II (WMO No.49), *Technical Regulations, Volume III Hydrology* (WMO No.49), loose-leaf, En, Fr, Ru, Es, editions 1971 and 1975; *Guide to Hydrometeorological Practices*, 2nd ed., 1970 (WMO No.168, T.P.82), loose-leaf, En, Fr, Es; *Guide to Hydrometeorological Practices*, 3rd ed., 1975 (WMO No.168), loose-leaf, En, Ru; *WMO Sea-Ice nomenclature* (WMO No.259); *Guide on the Global Data-processing System* (WMO No.305);
Manual on Codes (WMO No.306); *Manual on the Global Telecommunication System* (WMO No.386); *International Cloud Atlas* (WMO No.407); *Guide to Marine Meteorologi-*

cal Services (WMO No.471); *Guide on the Global Data-processing System* (WMO No.485); *Guide on the Global Observing System* (WMO No.480); *Manual on the Global Observing System* (WMO No.544)
Services: reference; loan
Operation:
computer-assisted
Data base description:
Size: 12. 000 terms (3000 in 4 languages)
Classification: UDC; alphabet
Terminological data:
definitions; equivalents
Data processing:
Hardware: IBM; 370/158; for multilingual applications a composite chain, including cyrillic characters is used
Software:
operation mode: batch; programming language(s): PL 1
Data available:
with charge

(Project(s))
Title: Revision of the International Meteorological Vocabulary
Project type: institutional; in progress
Principal investigator: Mr. Malcolm Rigby (USA)
Collaborators: 8 (part time)
Scope and/or subject matter: Meteorology and related fields. Languages covered: En, Fr, Ru, Es
Method of work: Terms and definitions are arranged by UDC and by alphabet
Publications: *International Meteorological Vocabulary*

Title: Revision of UDC Section 551. 5 and 556
Project type: institutional; in progress
Principal investigator: Mr. Malcolm Rigby (USA)
Collaborators: Work carried out by Working Group of the WMO 8 (working part time)
Scope and/or subject matter: Hydrology and related fields
Method of work: Terms and definitions are arranged by UDC and by alphabet
Publications: *International Glossary of Hydrology*

● ***310*** *(Data Bank)*
World Meteorological Organization/Organisation météorologique mondiale
WMO/OMM
41, avenue Giuseppe-Motta
No. 5 — Phone: (022) 34 64 00
Genéve 20, CH-1211 — Telex: 23260
Switzerland — Telegr.: Meteomond, Genève
Staff: 11
Head: Prof. A. C. Wiin-Nielsen, Secretary General
Expert(s): Mr. Malcolm Rigby (Chairman of the Working Group on Bibliographic Problems of the Commission for Atmospheric Sciences (CAS)), 5816 22nd St. N., Arlington, Va. 22205, USA
Staff categories:
1 translator(s) (part time)
Founded: 1951
Purpose of data bank/users: Production and updating of WMO publication No. 385 - *International Glossary of Hydrology*
Operation: computer-assisted
Subject(s): hydrology (and related fields)
UDC 556
Language(s): En; Fr; Ru; Es
Terminological data:
definitions; equivalents
Products: WMO Publication No. 385 - *International Glossary of Hydrology* (UDC 556: 030.8 = 00), ISBN 92-63-00385-8
Service charges
Data available:
with charge
Data processing:
Hardware/Capacity: IBM 370/158; For multilingual applications a composite chain, including cyrillic characters is used
Software:
operation mode: batch; programming language(s): PL 1

● ***311*** *(Organization)*
Conseil international de la langue française
[International Council of the French Language (not official)]
CILF
103, rue de Lille — Phone: 705. 07. 93
Paris, F-75007
France
Staff:
Head: M. Hubert Joly (Secrétaire général)
Expert(s): Mme Colette Murcia
Staff categories:
4 terminologist(s) (theory); 1 documentalist(s)/librarian(s)
Founded: 1967
Documentation:
bibliographic data
factographic data
Subject(s): all fields; linguistics
UDC 0/9
Language(s): Fr
Vocabularies:
Subject(s): all fields
UDC 0/9
Language(s): Fr
Training: In-house training
Publications: Reviews: *La Banque des mots* (semestrial); *La Clé des mots* (monthly until 1980); *Langues et terminologies* (bimonthly); *Le fellah* (trimonthly); *Le Français moderne* (trimonthly); Technical manuals for Third World countries; monolingual and bilingual works on languages with oral tradition; *Guide de la néologie*, 1981.
A large number of vocabularies have been published, such as: *Lexique photo-cinéma*, 1972; *Vocabulaire de la radiodiffusion*, 1972; *Langage médical moderne*, 1974; *Vocabulaire de la chasse et de la vénerie*, 1974; *Vocabulaire de la publicité*, 1976; *Vocabulaire de l'environnement*, 1976; *Dictionnaire forestiers multilingue*, 1976; *Vocabulaire de l'océanologie*, 1976; *Vocabulaire de l'écologie*, 1979; *Vocabulaire de la micrographie*, 1980; *Vocabulaire de l'astronomie*, 1980; *Vocabulaire de l'équitation et des courses*, 1981.
In addition, publications on a number of linguistic problems have been issued.
Services: reproduction; other: library services; annual colloquium; terminology bank.
Data processing:
Software:
retrieval documentation system: CILFOTERM - automatic terminological research index

(Project(s))
Title: Dictionnaire de mots scientifiques et techniques

● ***312*** *(Organization)*
General Agreement on Tariffs and Trade/Accord général sur les tarifs douaniers et le commerce
[Acuerdo general sobre aranceles aduaneros y comercio (Es)]
GATT
154, rue de Lausanne — Phone: 022 - 31 02 31
Genèva 21, CH-1211 — Telex: 28 787 GATT CH
Switzerland — Telegr.: GATT GENEVE
Staff: 1
Head: Mr. A. Dunkel, Director-General (as of 1 October 1980)
Expert(s): Mr. P. Y. Garaud, Director, Languages Division
Staff categories:
1 documentalist(s)/librarian(s)
Founded: 1948
Documentation: factographic data

Subject(s): trade; finance; economics
UDC 33; UDC 339
Language(s): En; Fr; Es
Vocabularies:
Subject(s): trade; finance; economics; commodities
UDC 33; UDC 339
Language(s): En; Fr; Es
Publications: *GATT Activities* (Info.booklet, annual, En, Fr, Es); *GATT What It Is* (Info.booklet, occasional, En, Fr, Es); *Basic Instruments and Selected Documents* (annual, En, Fr, Es; 30 issues 1952–1979); *International Trade* (annual report, En, Fr, Es); *Survey of Developments in Trade Policy* (tri-yearly, En, Fr); *Studies in International Trade* (occasional, En, Fr, Es; 8 issues)
Glossaire anglais-français à l'usage des traducteurs du GATT/English-French glossary for the use of GATT translators/Glosario inglés-francés para uso de los traductores de GATT; Glossaire anglais-espagnol á l'usage des traducteurs du GATT/English-Spanish glossary for the use of GATT translators/Glosario inglés-español para uso de los traductores del GATT; Glossaire de termes concernant le Centre du Commerce International CNUCED/GATT/Glossary of terms concerning the International Trade Centre UNCTAD/GATT/Glosario de términos relativos al Centro de Comercio Internacional UNCTAD/GATT; Supplement au glossaire de termes concernant le Centre du Commerce International CNUCED/GATT/Supplement to glossary of terms concerning the International Trade Centre UNCTAD/GATT/Suplemento del Glosario de términos relativos al Centro de Comercio Internacional UNCTAD/GATT;
Liste de termes de la passation des marchés publics/List of terms on government procurement/Vocabulario de las compras del sector público; Liste de termes concernant les produits laitiers/List of terms concerning dairy products/Vocabulario relativo a los productos lácteos;
Glossaire de la sellerie-bourrellerie/Glossary on saddlery and harnessmaker's wares/Vocabulario de guarniciones y otros articulos para caballerias; Liste de termes concernant les bois tropicaux et leurs usages/List of terms concerning tropical timber and uses thereof/Vocabulario de maderas tropicales y términos conexos;
Terminologie de l'inflation/Terminology of inflation/Terminologia de la inflación;
Liste de termes d'ésthetique industrielle/List of terms on industrial design/Vocabulario de diseño industrial;
Liste de termes relatifs á l'imposition des societés/List of terms on company taxation/Vocabulario de la tributación de las sociedades; Glossaire des questions douanières/Glossary of customs tariff matters/Vocabulario de cuestiones aduaneras;
Liste de termes techniques de l'industrie textile/List of technical terms for the textile industry/Vocabulario de la industria textil;
Liste de termes relatifs á l'arrangement concernant le commerce international des textiles/List of terms concerning the arrangement regarding international trade in textiles/Vocabulario del acuerdo relativo al comercio internacional de los textiles;
Glossaire de la peausserie-maroquinerie/Glossary of the hides, skins and leather goods sector/Vocabulario del cuero, la piel y la marroquineria; Liste de termes relatifs aux négociations sur les tarifs douaniers/List of terms concerning the negotiations on customs tariffs/Terminologia de las negociaciones arancelarias
Services: reference
No service charges
Operation:
manual
Data base description:
Name: Terminology File
Size: Unevaluated
Language(s): En; Fr; Es
Classification: Alphabetical
Terminological data:
definitions; explanations; context; equivalents
Access:
via telephone/via telex
Data available:
on exchange basis

◐ *313* *(Organization)*
Comité européen des constructeurs de machines à bois
EUMABOIS
150, Boulevard Bineau
Neuilly-sur-Seine, F-92000 Telex: 610460F
France Telegr.: SYMOFRA - Neuilly-sur-Seine
Staff:
Head: M. J. Chaumet
Founded: 1960
Documentation: bibliographic data
Subject(s): woodworking machines
UDC 674. 05
Language(s): Fr; En; De

● *314* *(Commission/Committee)*
Commission formation de la FEANI
[Formation Commission; Ausbildungskommission (De)]
Parent organization: Féderation européenne d'associations nationales d'ingénieurs (FEANI); (20 countries)
4 rue de la Mission Marchand Phone: 224 91 43
Paris, F-75016
France
Contact: Prof. Ugo Mocafico
Institut de machines hydrauliques
Ecole polytechnique fédérale
26 av. de Cour
CH-1007 Lausanne
Committee members: 12
Chairperson: Prof. Ugo Mocafico
Established: 1970
Vocabularies:
Subject(s): engineering professions; curriculum
UDC 62; UDC 372
Language(s): En; Fr; De
Vocabulary standards:
Subject(s): engineering professions; curriculum
UDC 62; UDC 372
Language(s): En; Fr; De
Publications: Work now in progress in cooperation with WFEO.

(Project(s))
Responsible: Commission formation de la FEANI
Title: Glossary of technical terms in the field of engineering professions
Principal investigator: Prof. Mocafico
Collaborators: 3
Scope and/or subject matter: To establish a glossary of technical terms related to the engineering profession and education, starting from the French language and giving definitions and possible equivalent terms in English and German; extension to other languages is also considered.
Method of work: Circulation between WFEO-members (essentially the French speaking countries as Algeria, Canada etc.) for coordination

● *315* *(Organization)*
Commission of the European Communities. Bureau of Terminology/Commission des Communautés européennes. Bureau de terminologie
CEC - BTL/CCE - BTL
Rue Alcide de Gasperi, Bâtiment Jean Monnet A2–100
P. O. Box 1907 Phone: 4301–2389
Luxembourg, L-1615 Telex: 3423 or 3446 COMEU LU
Luxembourg
Staff: 39
Expert(s): Mr. Jacques Goetschalckx
Staff categories:
24 terminologist(s) (theory); 5 documentalist(s)/librarian(s); 10 other professional(s)
Founded: 1963
Documentation:
bibliographic data
factographic data
Subject(s): all fields
UDC 0/9

Language(s): De, Fr, It, Nl, Da, En; Gr; Pt; Es
Vocabularies:
Subject(s): all fields
UDC 0/9
Language(s): De; Fr; It; Nl; Da; En; Gr; Pt; Es
Training: undergraduate; graduate; postgraduate
Access to training: available to outside users
Curriculum: lexicographical methods
Publications: *Glossaire Congrès II - L'acier et l'esthétique industrielle*, 1965, *Progrès dans la transformation de l'acier*, 1965, 5 vol. (partiellement épuisé); *Glossaire Concours d'architectes - L'acier dans le bâtiment*, 1966, 5 vol.; *Glossaire Congrès III - L'acier dans l'agriculture* - 1966, 5 vol.;
Articulation de l'organigramme des services de la Commission des Communautés européennes (doc.1309/72); Liste des professions - Industrie charbonnière (doc.3339/68); Glossaire de base *Crédit et Investissements* (doc.3617/69); Cours d'initiation à l'exploitation des mines avec glossaire phraséologique en 5 langues (doc.3551/1/68); Projet de glossaire multilingue de l'informatique (doc.3403/69);
Glossaire sur *Les rayonnements X parasites émis par des composants électriques* (Colloque Toulouse, novembre 1970) (doc.3027/70) - (édition révisée 1974); Glossaire de *radioécologie* (Colloque Rome, septembre 1971)(doc.1871/71); Glossaire *Télécommande et télécommunications dans les charbonnages* (doc.1323/72); Glossaire sur *Le plomb dans l'environnement* (doc.2612/72); Glossaire sur *La terminologie de base de la documentation* (doc.1666/73);
Glossaire *Pesticides* (doc.1491/74); Glossaire *Routes* (édition révisée et élargie) (doc.3064/78); Glossary for the Congress on *Medical Aspects of Driving Accidents* (édition limitée, révision en cours); Glossaire *Sidérurgie* (terminologie de base) (doc.3393/77); Glossaire *Construction* (doc.3135/74);
Glossaire *Statistique* (doc.1412/75); Glossaire *Marchés publics* (dispositions communautaires) (doc.2732/75); Glossaire *Téléinformatique* (doc.3988/77); Glossaire *Eurofice* (doc.1453/77); Glossaire *Linguistique* - Traitement de l'information et I &D (édition limitée, révision en cours); Glossaire *Documentation* - Difficultés de terminologie et de traduction (doc.2855/79);
Liste terminologique pour les Journées d'information. *Nouveaux procédés de valorisation du charbon* (doc.4891/79); *Onshore/Offshore Oil and Gas Multilingual Glossary*, 1979 (doc.EUR 6554); Glossaire *Radioactive Waste* (doc.3219/80); Glossaire *Energy from Biomass* (édition limitée); *Système de classification du Dr. Lenoch* (rev.ed.); Liste de Livres BTL (doc.3307/75).
A complete listing of publications is available in *Circulaire cumulative*, published irregularly
Services: reference; referral
No service charges
User aids: Bulletin du Bureau de terminologie (Nos.1–45; available as from No.20)
Operation:
manual
computer-assisted
Data base description:
Name: EURODICAUTOM
Size: 130. 000 cards
Language(s): Official languages of the EC; at present: De, Fr, It, Nl, Da, En
Classification: Lenoch classification
Terminological data:
definitions; explanations; context; equivalents; other data elements
Data processing:
Hardware: Siemens; 7. 760; 3 Mb; 2. 800 Mb; 1 line printer; TTY terminals, page mode Siemens terminals (TCV 450 Olivetti, etc.)
Software:
operating system: BS 2000; operation mode: All modes; programming language(s): ASSEMBLER; COBOL; PL 1; retrieval documentation system: Eurodicautom 2, home-made software
Access:
on-line

(Project(s))
Title: Glossaire "Datenschutz, Datensicherung, Privatsphäre und Systemzuverlässigkeit" - "Data Protection, Data Security, Privacy, Systems Reliability - Protection des données, sécurité des données, vie privée, fiabilité des systèmes"

Title: Glossaire "Monde du travail" (terminologie syndicale) [Glossary "World of Labour" (Terminology of Unions)]

Title: Glossaire "Multilinguisme" (édition révisée et élargie du Glossaire Linguistique - Traitement de l'information et I & D)

Title: Glossaire "Télématique" [Glossary "Telematics"]

Title: Glossaire sur les déchets et leur recyclage [Glossary on Waste and its Recycling]

Title: Glossaire Types of Documents - Types de documents - Dokumentenarten

Title: La constitution d'un fichier sur la construction (complété continuellement) [Establishment of a data file on construction (updated continually)]

Title: Recueil terminologique "Fours à coke" [Terminological collection "Coke furnaces"]

Title: Recueil terminologique "Fours électriques" [Terminological collection "Electric furnaces"]

Title: Recueil terminologique "Hauts-fourneaux" [Terminological collection "Blast furnaces"]

Title: Série de recueils terminologiques plurilingues des divers domaines de la sidérurgie [A series of multilingual collections in the various fields of metallurgy of iron]

Title: Vocabulaire des "Thésauri" [Vocabulary of "Thesauri"]

● ***316*** *(Data Bank)*
Commission of the European Communities. Bureau of Terminology/Commission des Communautés europé ennes. Bureau de terminologie
CEC - BTL/CCE-BTL
Rue Alcide de Gasperi, Bâtiment Jean Monnet
P. O. Box 1907 Phone: 4301–2389
Luxembourg, L-1615 Telex: 3423 or 3446 COMEU LU
Luxembourg
Staff: 39
Expert(s): Mr. Jacques Goetschalckx
Staff categories:
24 terminologist(s) (theory); 10 other professional(s)
Purpose of data bank/users: Help inhouse translators; EC translators, EURONET-DIANE users
Founded: 1975
Operation: computer-assisted

Data base description:
Name: EURODICAUTOM
Size: Approx. 1. 7 million pairs of terms
Language(s): Official languages of EC, namely De, Fr, It, Nl, Da, En
Classification: Lenoch classification system
Subject(s): all fields
UDC 0/9
Language(s): Da; De; En; Fr; It; Nl; envisaged: Es; Pt; Gr
Terminological data:
definitions; explanations; classified layout (via classification); context; standardized terms; equivalents; other data elements
Products: Word lists for in-house users; magnetic tapes, disks(for exchange purposes)
No service charges
Access:
on-line (EURONET-DIANE)
Data available: Negotiable
Data processing:
Hardware/Capacity: Siemens 7. 760; 3 Mb; 2. 800 Mb; 2 line printers; Connections with some 25-30 European institutions and appr. 60 EURONET-DIANE users; TTY terminals; page mode Siemens terminals (TCV 450 Olivetti, etc.)
Software: BS 2000; operation mode: all; programming language(s): mainly Assembler, COBOL

● ***317*** *(Organization)*
Commission of the European Communities. The Brussels Terminology Office/Commission des Communautés européennes. Bureau de terminologie de Bruxelles/Kommission der Europäischen Gemeinschaften. Terminologiebüro Brüssel/Commissione delle Comunitá europee. Ufficio di Terminologia Bruxelles/Commissie van de Europese Gemeenschappen. Terminologiebureau Brussel/Kommissionen for de europaeiske Faellesskaber. Terminologiecentralen Bruxelles
CEC. BTB/CCE. BTB
200, rue de la Loi
Bruxelles, B-1049
Belgium
Phone: 02/736. 60. 00
Telex: 21877 COMEU B
Telegr.: COMEUR Bruxelles
Visiting address: 1, Avenue de Cortenbergh, Office 1-95 (1st floor), Brussels
Staff: 34
Head: Dr. H. Kowalski
Expert(s): Dr. R. Goffin (responsible for terminological data bank EURODICAUTOM); Ms. B. Qvistgaard (responsible for glossaries); Ms. N. Kotowski (responsible for research and information); Ms. H. De Bruyne (responsible for documentation, library)
Staff categories:
1 terminologist(s) (special field); 17 terminologist(s) (theory); 2 documentalist(s)/librarian(s); 14 other professional(s)
Founded: 1958
Documentation:
bibliographic data
factographic data
Subject(s): all fields
UDC 0/9
Language(s): De; En; Fr; It; Nl; Da; Gr; (Es and Pt are envisaged)
Vocabularies:
Subject(s): all fields (covering the main activities of the CEC)
UDC 0/9
Language(s): De; En; Fr; It; Nl; Da; Gr; (Es and Pt are envisaged)
Training: graduate; postgraduate
Access to training: Training possibilities exist for graduates from universities and special language schools to work on terminology and lexicography (five months training periods).

Publications: For a complete list of the considerable number of BTB-glossaries in print, see *Circulaire cumulative* (Terminological publications and programmes of the EC Institutions), issued once a year. Since 1962 the *Bulletin de la traduction* is published 3 5 times a year (with terminological, linguistic and documentary information)
Services: Immediate supply of terminological, documentary and bibliographical information upon request from EC translators and other EC officials. Compilation and publication of glossaries and vocabularies
No service charges (for EC officials)
User aids: *Circulaire cumulative; Bulletin de la traduction*
Operation:
manual
computer-assisted
Data base description:
Name: EURODICAUTOM
Size: 160.000 terminological entities (from 2-6 languages)
Language(s): Fr; En; Da; De; It; Nl
Classification: home made classification system; automatic indexing method
Data processing:
Hardware: SIEMENS; 7760; 3M6; magnetic tape; line printers VDU (Siemens 8161); SIEMENS 8161; OLIVETTI TCV450; TTY terminals
Software:
operating system: BS 2000; operation mode: batch; real time; programming language(s): PL1; COBOL; ASSEMBLER; retrieval documentation system: Home made retrieval system
Access:
via telephone
on-line
Data available:
on exchange basis

(Project(s))
Title: Comités (communautaires) II [Committees (of the Communities)]
Scope and/or subject matter: Fr; En; It; Nl; Da; Fr; En; It; Nl; Da

Title: Vocabulaire de la bourse
[Vocabulary of Stock Exchange]
Scope and/or subject matter: Fr; En; It; Nl; Da

Title: Schola europaea (Vocabulaire de l'Organisation et des structures administratives) [Schola europaea (Vocabulary of the Organization and of the administrative structures)]
Scope and/or subject matter: Fr; En; De; It; Nl; Da

Title: Supplément espagnol au vocabulaire budgétaire [Spanish supplement to the Budgetary Vocabulary]

Title: Supplément espagnol, portugais au Vocabulaire economie-finance-monnaie [Spanish and Portuguese supplements to the Vocabulary of Economics-Finances-Money]

Title: Supplément espagnol, portugais, grec au Vocabulaire TVA [Spanish, Portuguese and Greek supplements to the VAT Vocabulary]

Title: Supplément et mise á jour du Vocabulaire budgétaire

Title: Supplément grec au Vocabulaire douanier [Greek supplement to the Customs Vocabulary]

Title: Supplément portugais aux vocabulaires budgétaire, douanier, tabac et vinification [Portuguese supplement to the Vocabularies of Budgets, Customs, Tobaccoes and Vinification]

Title: Terminologie budgétaire nationale [National Budget Terminology]

Title: Terminologie de la Sécurité sociale des travailleurs migrants (en colaboration avec l'équipe de traduction CA.SS.TM9) [Terminology of the social security of migratory workers (in collaboration with the translation team CA.SS.TM)]
Scope and/or subject matter: Fr; En; It; Nl; Da

Title: Vocabulaire céréales [Vocabulary of Cereals]
Scope and/or subject matter: Fr; En; De; It; Nl; Da

Title: Vocabulaire de la viticulture [Vocabulary of Viticulture]
Scope and/or subject matter: Fr; En; It; Nl; Da

Title: Vocabulaire des impôts indirects [Vocabulary of Indirect Taxes]
Scope and/or subject matter: Fr; En; It; Nl; Da

Title: Vocabulaire maladies de la civilisation [Vocabulary of Civilizational Diseases]

Title: Vocabulaire nouvelles sources d'énergie [Vocabulary of New Sources of Energy]

Title: Vocabulaire produits de la pêche [Vocabulary of Fishing Products]

● ***318*** *(Data Bank)*
Commission of the European Communities. The Brussels Terminology Office
CEC. BTB
200, rue de la loi — Phone: 02/736. 60. 00
Bruxelles, B-1049 — Telex: 21877 COMEU B
Belgium — Telegr.: COMEUR Bruxelles
Visiting address: 1, Avenue de Cortenbergh, Brussels
Purpose of data bank/users: To assist translations; prepare glossaries
Data base description: EURODICAUTOM is available, developed and fed in conjunction with the CEC terminology office in Luxembourg
Name: EURODICAUTOM
Size: Approx. 1. 7 million pairs of terms
Language(s): Official languages of EC, namely De, Fr, It, Nl, Da, En, (Es, Pt and Gr to be added).
Subject(s): all fields
UDC 0/9
Language(s): En; Fr; It; Nl; Da; Pt; (Es and Gr are envisaged)
Terminological data:
explanations; context; standardized terms; equivalents; other data elements
Products: Word lists for in-house users; glossaries
Data processing:
Hardware/Capacity: SIEMENS 7760; 3 Mb; magnetic tape; Line printers VDU (Siemens 8161)
Software: BS 2000; operation mode: batch; real time; programming language(s): Assembler, COBOL

◐ ***319*** *(Organization)*
Conférence européenne des ministres des transports
[European Conference of Ministers of Transport]
CEMT; ECMT
19, Rue de Franqueville — Phone: 524. 82. 00
Paris Cedex 16, F-75775 — Telex: 611040
France
Staff: 15
Head: M. Georges Billet, Secretaire Général
Expert(s): Mme. P. Coquand
Staff categories:
1 documentalist(s)/librarian(s); 1 information specialist(s)
Founded: 1953
Documentation: bibliographic data
Subject(s): transport
UDC 625. 7/. 8
Language(s): Fr; En; De
Vocabularies:
Subject(s): transport
UDC 625. 7/. 8
Language(s): Fr; En; De
Publications: CIDET/ICTED. *List of descriptors in the field of transport economics documentation/Liste de descripteurs d'économie des transports/ Deskriptorenliste der Verkehrswirtschaft.* Paris: CEMT, 1978, 121 p., A4
Operation:
computer-assisted
Data base description:
Name: Transdoc
Size: 4000
Language(s): Fr; En; De
Classification: Thesaurus (List of descriptors in the field of transport)
Terminological data:
explanations; equivalents
Data processing:
Hardware: Burroughs; 6800; printer; COM; Terminal
Software:
operating system: Autodoc; operation mode: real time; programming language(s): ALGOL

● ***320*** *(Organization)*
European Parliament/Parlement européen/Parlamento Europeo (It)/**Europäisches Parlament** (De)/**Europees Parlement** (Nl)/**Europa-Parlamentet** (Da)
EP/PE/PE(It)/**EP**(De)/**EP**(Nl)/**EP**(Da)
Centre européen, Plateau du Kirchberg
P. O. Box 1601 — Phone: 4300–1
Luxembourg — Telex: 3494 EUPARL LU /
Luxembourg (Grand-Duché) — 2894 EUPARL LU
Visiting address: Bureau I/1212 (Bâtiment Tour)
Staff: 18
Head: Mr. B. Constantini
Expert(s): Mr. A. Rohaert
Staff categories:
11 terminologist(s) (theory); 7 other professional(s)
Founded: 1965
Documentation:
bibliographic data
factographic data
Subject(s): law; politics; agriculture; environment; social sciences; economics; transport; medicine; taxes; education; energy
UDC 34; UDC 32; UDC 63; UDC 574; UDC 3; UDC 34; UDC 628. 5; UDC 61; UDC 336. 2; UDC 620. 9
Language(s): Fr; It; En; De; Nl; Da; Gr
Vocabularies:
Subject(s): law; politics; agriculture; environment; social sciences; economics; transport; medicine; taxes; education; energy
UDC 34; UDC 32; UDC 63; UDC 574; UDC 3; UDC 34; UDC 628. 5; UDC 61; UDC 336. 2; UDC 620. 9
Language(s): Da; Nl; En; Fr; De; It; Gr
Publications: *Terminology of the Rules of Procedure of the European Parliament*, 1973, 1 PE 32. 566, 185 p. (in Fr, It, En, De, Nl, Da), (out of print.New edition, incl.He, in preparation); *Terminology of the European Patent*, definitive edition, 1981, PE 64. 906, 527 p. (in Fr, It, En, De, Nl, Da); *Fiscal Terminology*, 1975, with Annex: List of taxes and duties levied in the Member States, PE 40. 809, 253 p. (in Fr, It, En, De, Nl; annex also in Da); *Terminology*

of the Fisheries Sector, 3rd ed., 1977, with Annex: List of fish, crustaceans and molluscs, PE 48. 787/rev., 170 p. (in Fr, It, En, De, Nl, Da; annex also in La);
Terminology of Social Security, 1974, PE 37. 760, 240 p. (in Fr, It, En, De, Nl, Da); *Economic Terminology*, 3rd ed., 1979, PE 55. 733, 451 p. (in Fr, It, En, De, Nl, Da); *Terminology of Human Rights*, 1976, PE 43. 330, 322 p. (in Fr, It, En, De, De, Nl, Da); *Glossary of Financial and Monetary Terms*, 3rd ed., 1980, PE 59. 904, 348 p. (in Fr, It, En, De, Nl, Da);
Terminology of Drugs and Narcotics, 2nd edition, 1978, PE 48. 166, 275 p. (in Fr, It, En, De, Nl, Da); *Terminology of the European Social Charter*, 1975, PE 38. 898, 84 p. (in Fr, It, En, De, Nl, Da); *Terminology of the ACP-EEC Association*, 1977, with Annex: List of ACP States, PE 499. 114, 448 p. (in Fr, It, En, De, Nl, Da); *Terminology of the European Company* (based on the third Proposed Statute of the Commission), 1981, PE 68. 367, 130 p. (in Fr, It, En, De, Nl, Da);
Terminology of a Preliminary Draft Convention for a European Trade Mark, 2 nd ed., 1981, PE 64. 907, 305 p. (in Fr, It, En, De, Nl, Da); *Terminology of the Environment*, 1974, + Danish Supplement, 1976 (Beiträge zur Umweltgestaltung, Heft B 9.Published by Erich Schmidt Verlag, Berlin), (in Fr, It, En, De, Nl, Da). Obtainable only from bookshops;
Legal Terminology of the European Communities, 1976, PE 42. 622, (2 vols.: Corpus, 564 p.and Index, 309 p.), (in Fr, It, En, De, Nl, Da); *Terminology of Town and Country Planning*, 1978, PE 49. 659, 382 p. (in Fr, It, En, De, Nl, Da); *Terminology of Elections*, 1979, PE 55. 568, 232 p. (in Fr, It, En, De, Nl, Da);
Parliamentary Forms (of the European Parliament), 1972, PE 32. 011, (in Fr, It, En, De, Nl, Da); *Governments of the Member States of the European Community. Composition and responsibilities.* (Loose-leaf edition with file. Updates are supplied free of charge), PE 34. 239 (in Da, De, En, Fr, It, Nl, He); *Terminological Information.* Annual publication (23 numbers issued up to date). Nos 18 (1975) to 23 (1981) incl. are still available.
Services: reference; loan; reproduction
No service charges
User aids: Terminological Information (annual publication), 23 numbers up to present, nos. 18 to 23 are still available

(Project(s))
Title: Partie danoise de la Terminologie fiscale [Danish part of Fiscal Terminology]
Project type: institutional; in progress

Title: Terminologie administrative, parlementaire, politique et des traités [Terminology of Administration, Parliament, Politics and Treaties]
Project type: institutional; in planning
Publications: Documentation and card index available

Title: Terminologie de l'éducation [Terminology of Education]
Project type: institutional; in planning
Publications: Documentation and card index available

Title: Terminologie de l'environnement (éd. révisée) [Terminology of the Environment (revised ed.)]
Project type: institutional; in progress

Title: Terminologie de l'Université européenne [Terminology of the European University]
Project type: institutional; in planning
Publications: Documentation and card index available

Title: Terminologie des Constitutions des pays-membres [Terminology of the Constitutions of the Member Countries]
Project type: institutional; in planning
Publications: Documentation and card index available

Title: Terminologie des nouvelles sources d'énergie [Terminology of New Sources of Energy]
Project type: institutional; in progress

Title: Terminologie forestière [Terminology of Forestry]
Project type: institutional; in progress

Title: Terminologie parlementaire anglaise [English Parliamentary Terminology]
Project type: institutional; in progress

○ ***321*** *(Organization)*
British Standards Institution
BSI
2 Park Street
London, W1A 2BS
U. K.
Phone: 44 1 629 9000
Telex: 266933 bsi
Telegr.: standards london w. 1.
Staff:
Head: Mr. D. G. Spickernell, Director General
Vocabulary standards:
Subject(s): all fields
UDC 0/9
Language(s): En

● ***322*** *(Organization)*
Centre for Computational Linguistics. University of Manchester Institute of Science and Technology
CCL-UMIST
Sackville Street
P. O. Box 88
Manchester, M60 1QD
UK
Phone: 061–236–3311 ext. 2258
Staff: (15 Professors; 3 Lecturers; 1 Research Fellow; 2 Research Assistants; 8 Postgraduates)
Expert(s): Prof. J. C. Sager
Staff categories:
1 terminologist(s) (special field); 1 linguist(s); 1 information specialist(s); 1 terminology teacher(s); 1 other professional(s)
Founded: 1824
Documentation:
bibliographic data
factographic data
Subject(s): paper science; audio-visual; engineering; data processing
UDC 62; UDC 659. 3; UDC 681. 3; UDC 676
Language(s): En; Fr; Gr; Es
Vocabularies: Glossaries composed by 2nd and 4th year students; theses by postgraduate students
Subject(s): science; technology
UDC 5/7
Language(s): En; Fr; Gr; Es
Training: undergraduate; graduate; postgraduate
Access to training: available to outside users
Curriculum: fundamentals of terminology; terminological principles; lexicographical methods. (The Department offers honours degree in Applied Language Studies and M.Sc.and Ph.D.by thesis.) The syllabus for the Applied and Computational Linguistics component of the B. Sc. in Applied Language Studies is: Yr. 1: General linguistics, mathematics for linguists, computing for linguistic applications I; Yr. 2: Contemporary linguistics I, statistical linguistics, computing for linguistic applications II, lexicography; Yr. 3: spent abroad (prescribed reading, study, reports); Yr. 4: Contemporary linguistics II, techniques of linguistic programming, terminology, introduction to sociolinguistics projects.
Publications: Sager, J. C.; Somers, H. L.; McNaught, J. *Guidelines for the Establishment of Comparison and Compatibility Matrices for Thesauri in the Social Sciences.*

(Report prepared for UNESCO) CCL report 81/2.
Sager, J. C.; McNaught, J. *Feasibility Study of the Establishment of a Terminological Data Bank in the U. K.* CCL report 81/8 and British Library R & D report 5642. Sager, J. C.; McNaught, J. *Selective Survey of Existing Linguistic Data Banks in Western Europe.* CCL reports 81/9 and British Library R & D report 5643. Price L7. 50 + p]; Sager, J. C.; McNaught, J. *Model Specification of a Terminological Data Bank for the U. K.* CCL report 81/10 and British Library report R & D 5644. Price L5. 50 + p]. All the above reports are only available from CCL
Services: consultancy; research & development; other: occasional research reports
Service charges: Cost of reproduction and mailing
Operation:
computer-assisted
Data processing:
Hardware: DEC-PDP; KARADAWN; INTERTEC DATA SYSTEMS; TANDY; Models: 11/03 (KD-2M; SUPERBRAIN (3); TRS-80); Memory: 64K; 128K; 64K; 64K; 1/2 MB on floppy disc drives; 2. 2 MB on floppy disc drives plus 32 MB on hard disc; 320K on floppy disc drives; 166K on floppy disc drives; VDU and LA 36 Decwriter; VDU & Olivetti Daisy Wheel printer; Centronics 730–2 printer; Epson TX-80 printer; Microline 80 printer
Software:
operating system: RT-11; CP/M; NEWDOS; UCSD Pascal system; programming language(s): Pascal (various dialects); C; Lisp; BCPL; Assembly languages

(Project(s))
Title: A contrastive semantic analysis of noun compound terms in English, French, and Spanish, with reference to the standardized terminology of the field of data processing
Project type: thesis, dissertation ; in progress
Principal investigator: Prof. J. C. Sager
Collaborators: 3
Scope and/or subject matter: The exploration of semantic relations expressed in noun concepts in three languages

Title: Analysis of word ambiguity and methods for the discovery of pronoun reference in machine translation
Project type: thesis, dissertation; in progress
Principal investigator: Prof. J. C. Sager; Mr. R. L. Johnson
Collaborators: 3
Scope and/or subject matter: Machine translation

Title: BEDE
Project type: institutional; in progress
Principal investigator: Mr. R. L. Johnson; Mr. H. L. Somers
Collaborators: Numerous
Scope and/or subject matter: Limited syntax machine translation system based on Interlingua, to translate texts in paper science and electronics

Title: Establishment of computational characteristics for an Integrated Descriptor Bank in the Social Sciences
Project type: commissioned project (UNESCO); in progress
Principal investigator: Prof. J. C. Sager
Collaborators: 4
Scope and/or subject matter: Machine readable thesauri in the social sciences

Title: In-depth study of machine translation systems
Project type: commissioned project (Science and Engineering Research Council and Social Sciences Research Council); in progress
Principal investigator: Prof. J. C. Sager
Collaborators: 6
Method of work: Collection and analysis of documentation and other material pertaining to MT systems

Title: Investigation into the syntactic structures of abstracts, and the definition of an interlingua for their translation by machine
Project type: thesis, dissertation; in progress
Principal investigator: Prof. J. C. Sager; Mr. R. L. Johnson
Collaborators: 3
Scope and/or subject matter: Machine translation of abstracts

Title: Microprocessor software development for a controlled vocabulary limited syntax machine translation system
Project type: thesis, dissertation; in progress
Principal investigator: Mr. R. L. Johnson; Mr. H. L. Somers
Collaborators: 2
Scope and/or subject matter: Machine translation

Title: Reduction of special language compound terms
Project type: thesis, dissertation; in planning
Principal investigator: Prof. J. C. Sager
Collaborators: 3
Scope and/or subject matter: terminology

Title: Terminological relationships - their role in encoding and accessing systematic multilingual specialized dictionaries
Project type: commissioned project (British Library R & D Dept.; DES Research Fellowship); in progress
Principal investigator: Prof. J. C. Sager
Collaborators: 3
Scope and/or subject matter: It has been demonstrated for restricted fields of German technology that systematic German-English specialised computer dictionaries may be constructed and accessed using a system of terminological relationships. The study proposes to extend these techniques to other fields and languages, to test the wider validity of the techniques and, if necessary, to refine or remodel them, and to adapt the techniques for conversational use
Publications: McNaught, J.; Johnson, R. L. An Indicative Abstract of *The Application of Computers to the Production of Systematic, Multilingual Specialised Dictionaries and the Accessing of Information Systems* by M. L. Hann, 1978. Fachsprache (1980), no. 2, p. 54–60; McNaught, J. *Terminological Relationships and Machine Dictionaries.* (Paper presented at Handelshøjskolen i København Forskningsseminar V 'Fagsprog - Teori og Praxis', Rolighed, May 1981.) CCL report 81/6

● ***323*** *(Organization)*
School of Modern Languages, University of Bath
Claverton Down — Phone: 61244
Bath, BA2 7AY — Telex: 449097
UK
Staff: 20
Head: Prof. J. Coveney
Founded: 1970
Subject(s): management
UDC 65
Language(s): Ar; Fr; En; De; It; Ru; Es
Training: postgraduate
Publications: *Glossary of French and English Management Terms* (1972); *Glossary of German and English Management Terms* (1977); *Glossary of Spanish and English Management Terms* (1978)
Data processing:
Hardware: HONEYWELL; LEVEL 68; TELETYPE 43; CIFER
Software:
operating system: MULTICS; operation mode: real time; programming language(s): BASIC; FORTRAN; eventually PASCAL

● *324* *(Organization)*
Scott Polar Research Institute
Parent organization: University of Cambridge
Lensfield Road
Cambridge, CB2 1ER
UK
Phone: Cambridge (0223) 66499
Staff: 24 (for entire institute)
Head: Dr. Gordon de Q. Robin
Expert(s): Dr. Terence Armstrong
Staff categories:
1 documentalist(s)/librarian(s); 1 information specialist(s), 2 translator(s)
Founded: 1920
Documentation:
bibliographic data
factographic data
Subject(s): snow; ice; glaciology
UDC 551. 326
Language(s): En; Fr; De; Es; Ru; Nl; Sv; No; Da; Is
Vocabularies:
Subject(s): sea; ice; snow
UDC 551. 326
Language(s): En
Vocabulary standards:
Subject(s): sea; ice; snow
UDC 551. 326
Language(s): En (with others given as equivalents)
Access to training: available to outside users
Publications: Armstrong, Terence, Roberts, B.; Swithinbank, Ch. *Illustrated glossary of snow and ice.* Cambridge: Scott Polar Research Institute, 1966, V, 60 p.
Services: reference; reproduction
Service charges
Data available:
free of charge

● *325* *(Commission/Committee)*
Scientific Committee on Antarctic Research, Working Group on Geology
SCAR WG G
Parent organization: Scott Polar Research Institute, Cambridge University
Lensfield Road
Cambridge, CB2 1ER
UK
Contact: Mr. R. J. Timpas
Secretary SCAR WG Geology
Bureau of Mineral Resources
P. O. Box 378
Canberra City
A. C. T. 2601
Australia
Committee members: 12
Chairperson: Dr. R. J. Adie
Established: 1961
Vocabularies:
Subject(s): antarctic geology
UDC 55
Language(s): En; Fr; Ja; Ru
Vocabulary standards:
Subject(s): stratigraphy ; rocks
UDC 55
Language(s): En

◐ *326* *(Data Bank)*
The British Library. Bibliographic Services Division
2 Sheraton Street
London, W1V 4BH
UK
Phone: 01–636 1544
Staff: 4
Expert(s): Mr. Derek Austin
Staff categories:
4 documentalist(s)/librarian(s)
Purpose of data bank/users: Generate "See" and "See also" references for PRECIS (Preserved Context Index System); Indexers
Operation: computer-assisted
Data base description:
Name: PRECIS RIN-file
Size: 70 000 terms
Language(s): En
Classification: PRECIS - thesaurus based on ISO 2788 and BS 5723
Subject(s): bibliography; indexing
UDC 016; UDC 025. 4
Language(s): En
Terminological data:
definitions (included where necessary); explanations (included where necessary); classified layout (indication of systematic relationships between concepts); context (only for homographs; other contexts handled in PRECIS strings); standardized terms
Products: Subject indexes (printed, fiche, etc.); microforms (RIN-file available on 9 fiches, 1: 42)
Data available:
with charge
Data processing:
Hardware/Capacity: IBM 370
Software:
operation mode: batch; programming language(s): Assembler; data base management system: OS/DOS

● *327* *(Organization)*
Transport and Road Research Laboratory, Department of the Environment and Transport
TRRL
Crowthorne, Berkshire,
RG 11 6AU
UK
Phone: Crowthorne 3131
Telex: 84 8 272
Staff: 600 (TRRL); 22 (Technical Information and Library Services)
Head: Mr. R. J. Bridle, Director of TRRL
Expert(s): Ms. P. E. Mongar, Head of Technical Information and Library Services
Staff categories:
15 documentalist(s)/librarian(s); 1 linguist(s); 15 information specialist(s); 2 translator(s)
Founded: 1930
Documentation: bibliographic data
Subject(s): road research; transport; urban planning
UDC 656; UDC 625. 7; UDC 711
Language(s): En; Fr; De
Vocabularies:
Subject(s): road; transport; urban planning
UDC 656; UDC 625. 7; UDC 711
Language(s): En; Fr; De
Publications: Unspecified number of publications
Services: reference; loan; translation (internal)
No service charges
Operation:
computer-assisted
Data processing:
Hardware: PRIME; 400

● *328* *(Organization)*
University of Exeter Language Centre
Queen's Building,
The Queen's Drive
Exeter, Devon, EX4 4QH
UK
Phone: 0392 - 77911 ext. 715
Staff: 21
Head: Dr. R. R. K. Hartmann
Expert(s): Dr. R. R. K. Hartmann
Staff categories:
5 linguist(s)
Founded: 1974
Documentation:
bibliographic data
factographic data
Subject(s): linguistics
UDC 801
Language(s): En; De

Vocabularies:
Subject(s): linguistics
UDC 801
Language(s): En
Theory: En
Training: undergraduate; postgraduate
Curriculum: lexicographical methods; occasional postgraduate essay, seminar, or dissertation
Publications: *Exeter Tapes.* A series of recordings for educational use by students in upper secondary, further and higher education, particularly in languages and literature, produced by Exeter University Language Centre under the general editorship of Keith Cameron and Reinhard Hartmann; over 300 tapes are currently available in a number of major and minor language areas covering a variety of topics including civilisation, language/linguistics, and literature. (A catalogue on the various topics is available on request from the Language Centre);
Exeter Linguistic Studies. Vol. I-V available; a complete list of the publications is available from the Language Centre
Services: translation (occasional); other: consultancy, (occasional)
Service charges
Operation:
computer-assisted
Data processing:
Hardware: I. C. L.; 4–72; 768Kb; 5 drives; various printers, including upper and lower case and high quality interchangeable head; approximately 100 teletype or similar
Software:
operating system: MJOB; operation mode: batch; interactive; programming language(s): BASIC; FORTRAN; SNOBOL 4

○ ***329*** *(Organization)*
Institute for Industrial Research and Standards
IIRS
Ballymun Road — Phone: 353 1 37 01 01 136
Dublin - 9 — 25449
Ireland — Telegr.: research, dublin
Head: Mr. Martin J. Cranley, Director General
Vocabulary standards:
Subject(s): all fields
UDC 0/9
Language(s): En

● ***330*** *(Organization)*
Bundesanstalt für Geowissenschaften und Rohstoffe
[Federal Institute for Geosciences and Natural Resources]
BGR
Stilleweg 2
P. O. Box 51 01 53 — Phone: 0511/6468–1
Hannover 51, D-3000 — Telex: 923 730 bfb(bgr) ha d
FRG — Telegr.: Geobund
Staff: 13
Head: Dipl.-Ing. Judit Nowak
Expert(s): Dr. B. König; Dr. G. Niedermeier
Staff categories:
7 documentalist(s)/librarian(s); 1 information specialist(s); 1 translator(s); 4 other professional(s)
Founded: 1969
Documentation: bibliographic data
Subject(s): geoscience; natural resources
UDC 55
Language(s): De; En; Fr
Publications: *Zentralblatt für Geologie und Paläontologie*, Teil I, Schweizerbart'sche Verlagsbuchhandlung, Stuttgart; *Bibliographie géologique et géophysique marines, zone méditerranéenne*, CIESM, Monaco
Services: SDI
Operation:
computerized
Data base description:
Name: Dokumentation der Geowissenschaften
Size: 250 000
Language(s): De
Classification: INIS -like
Terminological data:
standardized terms
Data processing:
Hardware: SIEMENS; 7541; 1000 kb; magnetic tape; other onsite peripherals; 2 Siemens Terminals
Software:
operation mode: batch; data base management system/retrieval documentation system: Siemens/Golem
Access:
via telephone / via telex
Data available:
with charge

● ***331*** *(Organization)*
Bundesanstalt für Straßenwesen
[Federal Institution for Road Research]
BAST
Brühler Straße 1
P. O. Box 510530 — Phone: 37021
Köln 51, D-5000 — Telex: 08882189 bas d
FRG
Head: Prof. Dipl. -Ing. P. Canisius
Founded: 1951
Documentation:
bibliographic data
factographic data
Subject(s): road research
UDC 625. 7/. 8; UDC 656. 1
Language(s): En; Fr; De
Vocabularies:
Subject(s): road research
UDC 625. 7/. 8; UDC 656. 1
Language(s): En; Fr; De
Services: reference; translation (internal)
Operation:
computerized
Data processing:
Hardware: SIEMENS; 7. 750
Software:
operating system: GOLEM

● ***332*** *(Organization)*
Bundessprachenamt
[Federal Office of Languages; Office fédéral des langues]
BSprA
Horbeller Straße 52 — Phone: 02233/55–1
Hürth, D-5030
FRG
Head: Mr. Hanns Maur, president
Staff categories:
15 terminologist(s) (theory)
Founded: 1969
Documentation:
bibliographic data
factographic data
Subject(s): technical fields
UDC 5/7
Language(s): De; En; Fr; Ru; It; Pt; Nl
Vocabulary standards:
Subject(s): technical fields
UDC 5/7
Language(s): De
Publications:
Word lists, glossaries, etc. for own needs
Services: translation
User aids: *Der Sprachmittler* (in-house publication); *Lebende Sprachen* (Co-Editor)
Operation:
computer-assisted
Data base description:
Name: Lexicographical Information System (LEXIS)
Size: 2 files of 1, 434, 000 entries each
Language(s): De; En; Fr; Ru; It; Pt; Nl
Terminological data:
definitions; explanations; context; equivalents; other data elements

Data processing:
Hardware: IBM; 3033; 3031; 8 Visual Display Units; 3 printers
Software:
operating system: IMS (Time Sharing); operation mode: batch; TP; interactive mode; programming language(s): PL/1; Assembler
Data available:
on exchange basis; with charge

● *333* *(Data Bank)*
Bundessprachenamt
[Federal Office of Languages; Office fédéral des langues]
BSprA
Horbeller Straße 52 Phone: 02233/55–1
Hürth, D-5030
FRG
Visiting address: As above
Head: Mr. Hanns Maur, president
Staff categories:
15 terminologist(s) (theory)
Founded: 1969
Purpose of data bank/users: Computer-assisted translation; Translators, interpreters, language teachers, experts in the field of information and documentation, standardization and technical writing
Operation: computer-assisted
Data base description:
Name: Lexicographical Information System (LEXIS)
Size: 2 files of 1, 434, 000 entries each
Language(s): De; En; Fr; Ru; It; Pt; Nl
Subject(s): technical fields
UDC 5/7
Language(s): De; En; Fr; Ru; It; Pt; Nl
Terminological data:
definitions; explanations; context; equivalents; other data elements
Products: Word lists, glossaries, etc. for own needs; microforms (COM)
Service charges
Data available:
on exchange basis; with charge
Data processing:
Hardware/Capacity: IBM 3033; 3031; 8 Visual Display Units; 3 printers
Software:
operating system: IMS (Time Sharing); operation mode: batch; TP; interactive mode; programming language(s): PL/1; Assembler

○ *334* *(Organization)*
DIN Deutsches Institut für Normung e. V.
[German Institute for Standardization]
DIN
Burggrafenstraße 4–10 Phone: 030 2601–1
P. O. Box 1107 Telex: 184273 din d
Berlin 30, D-1000 Telegr.: deutschnormen Berlin
FRG
Staff:
Head: Dr. H. Reihlen
Vocabulary standards:
Subject(s): all fields
UDC 0/9
Language(s): De

● *335* *(Commission/Committee)*
Arbeitsausschuß 1: Benennungsgrundsätze
[Terminological principles (not official)]
NAT-AA 1
Parent organization: DIN Deutsches Institut für Normung e. V., Normenausschuß Terminologie (NAT)
Burggrafenstraße 4–10
1107 Phone: 030/2601–318 or 2601–1
Berlin 30, D-1000 Telex: 184 273 din d
FRG Telegr.: Deutschnormen Berlin
Contact: Eva-Maria Baxmann, M. A; Normenausschuß Terminologie im DIN
Chairperson: Dr. E. Link
Vice-chairperson: O. Nüssler
Established: 1961
Vocabulary standards:
Subject(s): terminological principles
UDC 001. 4. 001; UDC 003. 62
Language(s): De
Publications: DIN 2330 *Begriffe und Benennungen; Allgemeine Grundsätze* (März 1979); DIN 2331 *Begriffssysteme und ihre Darstellung* (April 1980); DIN 2332 *Internationale Angleichung von Fachbegriffen und ihren Benennungen* (Februar 1979); DIN 2338 Teil 1: *Begriffssystem Zeichen; Allgemeine Grundlagen* (Entwurf Februar 1979); DIN 2338 Teil 2: *Begriffssystem Zeichen; Zeichentypologie* (Entwurf März 1980); DIN 2340 *Abkürzungsregeln für Benennungen von Fachbegriffen* (März 1977); DIN 31 620 *Abkürzungsregeln für Bezeichnungen in der juristischen Fachsprache* (Entwurf Juli 1976)

● *336* *(Commission/Committee)*
Arbeitsausschuß 2: Lexikographische Grundsätze
[Lexicographical principles (not official)]
NAT-AA2
Parent organization: DIN Deutsches Institut für Normung e. V., Normenausschuß Terminologie (NAT)
Burggrafenstraße 4–10
1107 Phone: 030/2601–318 or 2601–1
Berlin 30, D-1000 Telex: 184 273 din d
FRG Telegr.: Deutschnormen Berlin
Contact: Eva-Maria Baxmann, M. A; Normenausschuß Terminologie im DIN
Chairperson: Peter Port, M. A.
Vice-chairperson: Dr. E. Kahleyss; O. Vollnhals
Established: 1961
Vocabulary standards:
Subject(s): lexicographical principles
UDC 001. 4; UDC 801. 316
Language(s): De
Publications: DIN 2333 *Fachwörterbücher; Stufen der Ausarbeitung* (Vornorm September 1974); DIN 2335 *Terminologie und Lexikographie; Sprachenzeichen und ihre Ergänzung durch Länderzeichen und Autoritätszeichen* (Vornorm Januar 1976); DIN 2336 *Lexikographische Zeichen für manuell erstellte Fachwörterbücher* (März 1979); DIN 2339 *Ausarbeitung und Gestaltung von Publikationen mit terminologischen Festlegungen* (Entwurf Februar 1978)

● *337* *(Commission/Committee)*
Arbeitsausschuß 5: Datenverarbeitungsgerechte Terminologie und Lexikographie
[Computational aids for terminology and lexicography (not official)]
NAT-AA 5
Parent organization: DIN Deutsches Institut für Normung e. V., Normenausschuß Terminologie (NAT)
Burggrafenstraße 4–10
1107 Phone: 030/2601–318 or 2601–1
Berlin 30, D-1000 Telex: 184 273 din d
FRG Telegr.: Deutschnormen Berlin
Contact: Eva-Maria Baxmann, M. A., Normenausschuß Terminologie im DIN
Chairperson: J. A. Bachrach
Vice-chairperson: G. Beling
Established: 1961
Vocabulary standards:
Subject(s): computational aids for terminology and lexicography
UDC 001. 4; UDC 681. 3; UDC 801. 316
Language(s): De

Publications: DIN 2341 *Magnetband-Austauschformat für terminologische/lexikographische Daten* (MATER) (Vornorm Mai 1980)

● ***338*** *(Commission/Committee)*

DIN Deutsches Institut für Normung e. V., Normenausschuß Terminologie

[Technical Committee Terminology of DIN (not official)]

DIN-NAT

Parent organization: DIN Deutsches Institut für Normung e. V.

Burggrafenstraße 4–10
P. O. Box 1107
Berlin 30, D-1000
FRG

Phone: 030/2601–318 or 2601–1
Telex: 184 273 din d
Telegr.: Deutschnormen Berlin

Chairperson: Eva-Maria Baxmann, M. A. (Secretary)
Established: 1961
Vocabularies:
Subject(s): terminology; lexicography (vocabulary)
UDC 001. 4; UDC 801. 316
Language(s): De
Vocabulary standards:
Subject(s): terminological principles; lexicographical principles
UDC 001. 4. 001
Language(s): De
Publications: Monthly list of DIN-Standards containing standardized concepts or terms, published in *DIN-Mitteilungen*, ed. by DIN Deutsches Institut für Normung, Berlin. Articles and reviews in *DIN-Mitteilungen*

(Project(s))
Responsible: NAT/AA 5: Datenverarbeitungsgerechte Terminologie und Lexikographie
Title: Erarbeitung und Einführung IuD-spezifischer Normen: DV in der Terminologiearbeit [Elaboration and implementation of standards for information and documentation: Data processing for terminological purposes]
Principal investigator: Eva-Maria Baxmann, M. A. (Secretary)
Scope and/or subject matter: Elaboration and implementation of standards for information and documentation: Data processing for terminological purposes
Method of work: Standardization work
Publications: DIN 2341 (Preliminary Standard, May 1980)

Responsible: NAT/AA 1: Benennungsgrundsätze ; NAT/AA 2: Lexikographische Grundsätze
Title: Erarbeitung und Einführung IuD-spezifischer Normen: Grundlagen terminologischer Arbeit [Elaboration and implementation of standards on information and documentation: Terminological principles]
Principal investigator: Eva-Maria Baxmann, M. A. (Secretary)
Scope and/or subject matter: Elaboration and implementation of standards on information and documentation: Terminological principles
Method of work: Standardization work
Publications: DIN 2330 (March 1979); DIN 2331 (April 1980); DIN 2332 (February 1979); DIN 2335 (Preliminary Standard September 1976); DIN 2338 Part 1 (Draft February 1979); DIN 2338 Part 2 (Draft March 1980); DIN 2339 (Draft February 1978); DIN 2340 (March 1977); DIN 31 620 (Draft July 1976)

● ***339*** *(Data Bank)*

DIN Deutsches Institut für Normung e. V.

[DIN German Standards Institution (not official)]

DIN

Burggrafenstraße 4–10
P. O. Box 1107
Berlin 30, D-1000
FRG

Phone: 030/2601–318 or 2601–1
Telex: 184 273 din d
Telegr.: Deutschnormen Berlin

Head: Dr. Ing. Helmut Reihlen
Expert(s): Eva-Maria Baxmann, M. A., Normenausschuß Terminologie im DIN
Founded: 1917
Purpose of data bank/users: Documentation of standardized concepts (terms, definitions, synonyms, etc.) contained in DIN-Standards; Standards committees, translators, etc.
Operation:
computerized
Data base description:
Name: Terminologie-Datenbank des deutschen Normenwerks (TERM)
Language(s): De; En; Fr; and others
Subject(s): all technical fields
UDC 0/9
Language(s): De
Terminological data:
definitions; explanations; classified layout; standardized terms; equivalents; other data elements
Data available:
with charge

● ***340*** *(Organization)*

Fachhochschule für Bibliothekswesen Stuttgart

[Stuttgart Library School]

FHB

Feuerbacher Heide 38–42
Stuttgart 1, D-7000
FRG

Phone: 0711–221083/84

Staff: 50
Head: Prof. Dr. Hermann Wassner
Expert(s): Prof. Dipl. Ing. Bernhard Kirchhof
Founded: 1942 (1945/46)
Activities: Terminological activities focus on documentation for teaching purposes
Documentation: bibliographic data
Subject(s): documentation; librarianship; information
UDC 02
Language(s): De; En
Training: Reference to and practical use of vocabularies and thesauri during training courses
Services: loan; reproduction
No service charges
Access:
via telephone
Data available:
on exchange basis; with charge

● ***341*** *(Organization)*

Fachinformationszentrum "Hüttenkunde, Werkstoffe, Metallbe- und -verarbeitung"

[Information Centre "Metallurgy, Materials and Metalworking"]

FIZ 5; FIZ 5

Parent organization: Bundesanstalt für Materialprüfung

Unter den Eichen 87
Berlin 45, D-1000
FRG

Phone: (030) 8104–0050/1
Telex: 01–832–61 bamb d

Head: Dr. -Ing. P. Büttner
Expert(s): Dipl. Min. J. -G. Ilter
Staff categories:
information specialist(s) (unspecified)
Founded: 1871
Documentation:
bibliographic data
factographic data
Subject(s): metals; materials
UDC 669. 2/. 8
Language(s): De

(Project(s))
Title: Dreisprachiger Thesaurus Metallurgie, Hüttenkunde, Metallbe- und -verarbeitung [Trilingual Thesaurus on Metallurgy and Metalworking]
Project type: institutional; in progress
Collaborators: 3 to 6
Source material: Input for database SDIM 2

● *342* *(Organization)*
Freie Universität Berlin, Fachbereich Kommunikationswissenschaften, Arbeitsbereich Informations- und Dokumentationswissenschaft
[Free University Berlin, Dept. for Communication Sciences, Section for Information and Documentation Science (not official)]
Parent organization: Freie Universität Berlin
Hagenstraße 56 Phone: 030/8263006
Berlin 33, D-1000
FRG
Staff: 2 (research project: 4)
Head: Prof. Dr. Gernot Wersig
Expert(s): Prof. Dr. Gernot Wersig
Founded: 1971
Documentation:
bibliographic data
factographic data
Subject(s): information; documentation; information/communication technologies
UDC 659; UDC 002; UDC 007
Language(s): De; En
Vocabularies: Not established recently
Vocabulary standards:
Subject(s): information; documentation
UDC 007; UDC 002
Language(s): De; En
Theory: (in general)
Training: graduate; postgraduate
Access to training: in principle
Curriculum: fundamentals of terminology; terminological principles
Publications: Report series (*Informationswissenschaftliche Forschungsberichte aus der Freien Universität Berlin* [Research Reports on Information Science of the Free University of Berlin]; book series (*Beiträge zur Informations- und Dokumentationswissenschaft* [Contributions to Information and Documentation Science] Verlag Saur, München/New York/Paris/London)
Services: Scientific advice
Service charges: Negotiable

● *343* *(Organization)*
Hochschule Hildesheim - Studiengang Fachübersetzen
[Graduate course for technical translators at the Hochschule of Hildesheim]
Marienburger Platz Phone: 05121/81061
Hildesheim, D-3200
FRG
Staff: 9
Head: Prof. Dr. Reiner Arntz
Expert(s): Prof. Dr. Reiner Arntz
Staff categories:
1 linguist(s); 4 translator(s); 1 terminology teacher(s); 2 other professional(s)
Founded: 1979
Documentation: factographic data
Subject(s): science; technology
UDC 5/6
Language(s): De; En; Fr
Vocabularies:
Subject(s): technical engineering; mechanical engineering
UDC 62
Language(s): De; En; Fr
Theory: De
Training: graduate
Curriculum: fundamentals of terminology; terminological principles; lexicographical methods

◐ *344* *(Organization)*
Institut für Geschichte der Medizin
Parent organization: Bayerische Julius-Maximilians-Universität Würzburg
Koellikerstraße 6 R Phone: 0931/31716
Würzburg, D-8700
FRG
Staff: 6
Head: Prof. Gundolf Keil, M. D., Ph. D., B. Sc.
Expert(s): Helmut Winkler, M. A.; Peter Proff, Assessor
Staff categories:
1 documentalist(s)/librarian(s); 1 linguist(s); 2 translator(s)
Founded: 1957
Documentation: factographic data
Subject(s): medicine
UDC 61
Language(s): Gr; La; De; En; Fr
Theory: De
Training: undergraduate
Curriculum: fundamentals of terminology; terminological principles

● *345* *(Commission/Committee)*
Kommission Begriffsbestimmungen im Gasfach
[Concept definitions in the field of industrial gases (not official)]
Parent organization: DVGW Deutscher Verein des Gas- und Wasserfaches e. V., technisch-wissenschaftliche Vereinigung
Frankfurter Allee 27
5240 Phone: 06196/44059
Eschborn, D-6236 Telex: 417420 DVGW D
FRG
Chairperson: Dipl. Ing. G. Dette
Vocabulary standards:
Subject(s): gas technology
UDC 621. 6. 02
Language(s): De
Publications: *Lexikon der Gastechnik*. Essen: Vulkan-Verlag, 1978

● *346* *(Organization)*
Siemens AG, Sprachendienst
[Siemens AG, Language Services Dept. ; Siemens AG, Service de traduction]
TEAM (Terminologie-Erfassungs- und Auswertungsmethode); TEAM (Terminology Evaluation and Acquisition Method)
Hofmannstraße 51
P. O. Box 70 00 72 Phone: 8400 291 (DDD-code 089)
München 70, D-8000 Telex: 5288–0
FRG Telegr.: siemens nachr, munich
Visiting address: Landsberger Str. 33, D-8034 Germering, FRG
Staff: 235 (plus about 400 free-lance collaborators)
Head: Dr. Th. Schneider, Linguistic Projects Manager; Mr. E. Tanke
Expert(s): Mr. J. Schulz
Staff categories:
1 terminologist(s) (special field)/terminology teacher(s); 11 terminologist(s) (theory); 6 linguist(s); 1 information specialist(s); 181 translator(s); 35 other professional(s)
Founded: Company: 1847
Documentation:
bibliographic data
factographic data
Subject(s): natural sciences; technology; law; accounting etc.
UDC 3/7
Language(s): Ar; De; En; Fr; It; Nl; Pt; Ru; Es
Vocabularies:
Subject(s): natural sciences; technology; law; accounting
UDC 3/7
Language(s): Ar; De; En; Fr; It; Nl; Pt; Ru; Es
Vocabulary standards: Member of DIN Normenausschuß Terminologie/AA1 "Benennungsgrundsätze"; member of ISO/TC 37; member of Infoterm/TermNet working group "Vocabulary of terminological data banks"
Training: undergraduate
Access to training: on request
Curriculum: fundamentals of terminology; terminological principles; lexicographical methods
Publications: Brinkmann, K. H.; Tanke, E. *TEAM - ein*

Programmsystem für die Lösung terminologischer und lexikographischer Aufgaben. Siemens-Schriftenreihe *data praxis*, Siemens AG, Munich, No. D 14/4059 (1969/1972); Brinkmann, K. H.; Tanke, E. *TEAM - A program system for the solution of problems in terminology and lexicography.* Siemens publication series *data praxis*, Siemens AG, Munich, Order No. D 14/4059–101 (1971); Brinkmann, K. H.; Tanke, E. *TEAM - programmnaja systema dlja rešenija terminologičeskih i leksikografičeskih zadač.* Siemens AG, Munich (mimeographed 1970);

Brinkmann, K. H.; Schulz, J. ; Tanke, E. *Das Wörterbuch aus der Maschine.* data report 4 (1969), Heft 4 ; Brinkmann, K. H. ; Schulz, J. ; Tanke, E. *Le dictionnaire électronique.* data report 5 (1970), No. 1; Brinkmann, K. H.; Schulz, J.; Tanke, E. *Il dizionario con il calcolatore.* data report (1970), n. 3;

Schulz, J. *Le système TEAM, une aide à la traduction.* Actes du colloque international de linguistique et de traduction, Montréal, 30. 9. - 3. 10. 1970. META Journal des traducteurs, Vol. 15 (1971), No. 1–2;

Brinkmann, K. H.; Müller, E. P. *Computer fördern die Verständigung.* Almanach 1971. Köln: Carl Heymanns Verlag, 1971;

Tanke, E. *Das aktuelle Wörterbuch aus der Datenbank.* Ein Beitrag zur Lösung sprachlicher Probleme im internationalen Informationsaustausch. Deutscher Drucker Nr. 38 (1971);

Schulz, J. *Reversible Transliteration kyrillischer Buchstaben. Nachrichten für Dokumentation* 24 (1973), Nr. 6;

Brinkmann, K. H. *Überlegungen zum Aufbau und Betrieb von Terminologie-Datenbanken als Voraussetzung der maschinenunterstützten Übersetzung. Nachrichten für Dokumentation* 25 (1974; Nr.3;

Schulz, J. L*exikographie mit TEAM - Automatischer Satz von Wörterbüchern.*data report 9 (1974), Heft 1;

Schmidt, R.; Vollnhals, O. *Der Einsatz des lexikographischen Zweigs eines Datenbanksystems zur Herstellung eines phraseologischen Fachglossars. Lebende Sprachen XIX (1974), Heft 3–4;* Brinkmann, K. H. *Terminologists, Lexicographers, and Computers. Philips Terminology Bulletin, Vol. 3 (1974), No. 3/4;*

Tanke, E. *Electronic Data Processing in the Service of Translators, Terminologists and Lexicographers. Philips Terminology Bulletin, Vol. 4 (1975) No. 2/3;* Brinkmann, K. H. *The TEAM Program System. Philips Terminology Bulletin, Vol. 4 (1975) No. 2/3;* Tanke, E. *Future Developments. Philips Terminology Bulletin, Vol. 4 (1975) No. 2/3;* Schulz, J. *Der Computer hilft dem Übersetzer - Ein Verfahren zur Abfrage eines auf Magnetband gespeicherten mehrsprachigen Fachwörterbuches. Lebende Sprachen XX (1975) Heft 4;*

Schulz, J. *Automatische Abfrage einer Terminologie-Datenbank. Nachrichten für Dokumentation 27 (1976) Nr. 1;* Tanke, E. *Terminologie-Datenbank.* Bundesministerium für Forschung und Technologie. Forschungsbericht T 76–30, 1976; Tanke, E. *Siemens Linguistic Data Bank. Proceedings - Aslib Technical Translation Group, London, 1976;*

Göricke, H.; Schulz, J. *The Dictionary in the Computer - Possibilities of Directly Interrogating a Multilingual Terminology Data Bank via Video Display Units. Babel - Revue internationale de la traduction, Nr. 7 (1977);* Brinkmann, K. H. *Perspektiven der maschinellen Übersetzung.* Nachrichten für Dokumentation 29 (1978) Nr. 3;

Vollnhals, O. *Die datenbankgesteuerte Herstellung einer Wörterbuchreihe. Babel - Revue internationale de la traduction (Sept.1978);* Brinkmann, K. H. *Der Einsatz von Terminologiedatenbanken zur Lösung von Übersetzungsproblemen im fachsprachlichen Bereich. Fachsprache 1 (1979) Nr. 1–2;* Schulz, J. *Eine Terminologiedatenbank für Übersetzer. Philips Terminology Bulletin Vol. 8 (1979) No. 1/2;* Brinkmann, K. H. *Perspectives d'avenir de la traduction automatique. Meta - Journal des traducteurs Vol. 24 (1979) No. 3;* Tanke, E. H. *Implementing Machine Aids to Translation.* In: Barbara Snell, ed.: *Translating and the Computer.* North Holland Publishing Company, Amsterdam, *1979;*

Hohnhold, I. *Übersetzungsorientierte Terminologiearbeit - fachsprachlich orientiertes Übersetzen. Lebende Sprachen XXIV (1979) Heft 4;* Blaha, H. *Aufbau und Nutzungsmöglichkeiten einer Normen-Terminologie-Datenbank. Fachsprache 2 (1980) Nr. 4;* Brinkmann, K. H. *Present and Future Technical Aids to Easier Communication in Foreign Languages. Babel - Revue internationale de la traduction, No. 2 (1980);*

Schulz, J. A *Terminology Data Bank for Translators (TEAM). Meta - Journal des traducteurs Vol. 25, (1980) No. 2;* Brinkmann, K. H. *Quality Criteria for the Exchange of Terminological Data.* In: Infoterm Series 5 - Terminological Data Banks. K. G. Saur, München, 1980; etc.

Services: According to special agreement

Operation:

computer-assisted ((translation))
computerized (terminology)

Data base description:

Name: TEAM
Size: About 1 million terminological units (entries)
Language(s): Ar; De; En; Fr; It; Nl; Pt; Ru; Es
Classification: Adapted version of "Fachordnung Technik" of the German automotive, mechanical engineering and electrical engineering industries. Test version of a classification system tailored to the needs of the cooperating organizations

Terminological data:

definitions (but not in all entries); explanations (but not in all entries); context (but not in all entries); equivalents; other data elements (grammatical information, sources, references to systems, equipment, etc)

Data processing:

Hardware: SIEMENS; 7. 000 Series; Minimum 1 20 Kb; Capacity: Practically unlimited; magnetic tape (and disks); Laser printer; standard printers; COM-equipment; Digiset CRT-phototypesetting equipment; Video and other terminals, various types, with extended character sets
Software:
operating system: BS 1000; BS 2000; operation mode: batch and real time (interactive); programming language(s): Assembler, data base management system/retrieval documentation system: Non-standard (own development)

Access:

via telephone (via terminologist)/ via telex (via terminologist)
on-line (possible (operative within Department)); off-line

Data available:

on exchange basis; with charge; free of charge (according to special agreement)

(Project(s))

Title: Terminologie Subdatenbank TEAM (Arabisch) [Terminology sub-data bank TEAM (Arabic)]
Project type: commissioned project; in progress
Principal investigator: Mr. Raouf Hannalla
Collaborators: 2
Scope and/or subject matter: Arabic scientific and technical terminology
Method of work: Use and expansion of an existing terminological data bank system (viz.TEAM)
Source material: All available sources, especially the glossaries received through the courtesy of the Bureau for the Coordination of Arabization of the Arab League Educational, Scientific and Cultural Organization (ALECSO)

● ***347****(Data Bank)*

Siemens AG, Sprachendienst

[Siemens AG, Language Services Department]

TEAM (= Terminologie-Erfassungs- und Auswertungs-Methode); TEAM (= Terminology Evaluation and Acquisition Method)

Hofmannstraße 51
Postfach 700073
München 70, D-8000
FRG

Phone: (089) 84 00 290
Telex: 5 212 461
Telegr.: siemensnachr muenchen

Visiting address: Landsberger Straße 33, D-8034 Germering, FRG
Staff: Terminological data bank: 24, Language Services Dept.: 250
Head: Dr. Th. Schneider, Linguistic Projects Manager Mr. E. Tanke, Senior Engineer
Staff categories:
2 terminologist(s) (special field); 7 terminologist(s) (theory); 8 linguist(s); 2 terminology teacher(s); 3 other professional(s) (lexicographers)
Founded: 1967
Purpose of data bank/users: Machine-aided translation, educational, lexicographical applications; translators, terminologists, lexicographers, teachers, engineers, scientists etc.
Operation:
Data base description:
Name: TEAM
Size: Roughly 1 million terminological units
Language(s): Ar; En; Fr; De; Nl; It; Pt; Ru; Es
Classification: Originally the classification system of the Dokumentationsring der Elektroindustrie was adopted - now under modification
Subject(s): all fields (of science and technology, economics, accounting, etc.)
UDC 0/9
Language(s): Ar; En; Fr; De; Nl; It; Pt; Ru; Es
Terminological data:
definitions (as available); explanations (as available); classified layout (possible but not used at present); context (as available); standardized terms; equivalents (as available); other data elements (grammatical information, sources, references to systems, equipment, etc)
Products: Glossaries, word lists, text-synchronous lists etc.; microforms (microfiche, standard size, 1: 24, 1: 42, 1: 48); magnetic tapes, disks (with data bank output as specified by user). For a complete list of dictionaries produced by TEAM consult Siemens.
Services: Direct access (video terminals)
Service charges: As arranged with user
Access: After proper arrangements have been made via telephone / via telex
Data available:
on exchange basis; with charge
Data processing:
Hardware/Capacity: Siemens 7 000 series; Minimum: 250 Kb; unlimited; magnetic tape (and disks); Printers including laser printer, COM-equipment, CRT phototypesetting equipment, video terminals, printing terminals
Software: operating system: BS 1000, BS 2000; operation mode: batch, interactive (dialog); programming language(s): Assembler; data base management system/ retrieval documentation system: TEAM/TEAM

● ***348*** *(Organization)*
Universität Duisburg - Gesamthochschule FB 3 - Romanistik/Linguistik and Allgemeine Sprachwissenschaft
[University of Duisburg - Dept. of Romance Languages and General Linguistics; Université de Duisburg - U. E. R. de langues romanes et de linguistique générale]
Lotharstraße 65
P. O. Box 101629 — Phone: 0203/3052609
Duisburg 1, D-4100 — Telex: 855793 uni du d
FRG
Staff: 4
Head: Prof. Dr. Bernd Spillner
Staff categories:
3 linguist(s)
Founded: 1974
Documentation:
bibliographic data
factographic data
Subject(s): lexicology; French lexicology; French for special purposes; linguistic analysis (Fr-De)
UDC 801. 3; UDC 801. 316. 4; UDC 804. 0
Language(s): Fr; De
Theory: Fr; De
Training: graduate; postgraduate
Curriculum: fundamentals of terminology; terminological principles; lexicographical methods (French lexicology, French terminologies)
Publications: Spillner, B. *Normierungen in der französischen Lexik*, in: Rolf Kloepfer et al. (ed.) *Bildung und Ausbildung in der Romania.* Bd. II: *Sprachwissenschaft und Landeskunde.* München 1979, p. 213 - 238 ;
Spillner, B. *Die Normierung der französischen Sprache und ihre Konsequenzen in Bildung und Ausbildung*, ibid., p. 177 - 182; Spillner, B. *Termini und Sprachfunktionen in der literaturwissenschaftlichen Fachsprache*, in: Th. Bungarten (ed.) *Wissenschaftssprache. Beiträge zur Methodologie, theoretischen Fundierung und Deskription.* München 1981, p. 372 - 403;
Spillner, B. *Probleme der Syntax von Fachsprachen - an französischen Beispielen.* In: W. Kühlwein: A. Roasch (ed.) *Sprache: Lehren-Lernen.* Kongreßberichte der 11. Jahrestagung der Gesellschaft für Angewandte Linguistik, Bd. I. Tübingen, 1981, p. 41–48.
Operation:
manual
Data base description:
Name: Documents and publications (languages for special purposes)
Language(s): Fr; De
Terminological data:
equivalents
Data available:
on exchange basis

(Project(s))
Title: Fachsprachenforschung Französisch (Lexikologie und Terminologie) [Research in French languages for special purposes (lexicology and terminology)]
Project type: institutional; in progress
Principal investigator: Prof. Dr. Bernd Spillner
Collaborators: 2
Scope and/or subject matter: Terminological structures in sciences (physics, biology, chemistry, etc.)
Method of work: Analysis of French texts; comparison with standard language; contrastive analysis Fr - De

● ***349*** *(Commission/Committee)*
Scientific Commission on Rhetoric and Stylistics
[Commission scientifique "Rhétorique et Stylistique"]
Parent organization: AILA (Association internationale de linguistique appliquée)
Lotharstraße 65, University of Duisburg - FB 3
101629 — Phone: 0203/3052609
Duisburg 1, D-4100
FRG
Committee members: 100
Chairperson: Prof. Dr. Bernd Spillner
Established: 1978
Theory: En; Fr; De

◐ ***350*** *(Organization)*
Verein Deutscher Ingenieure
VDI
Graf-Recke-Straße 84, VDI-Haus
Postfach 11 39 — Phone: 0211/62 14–1
Düsseldorf 1, D-4000 — Telex: 8 586 525 vdi
FRG — Telegr.: ingenieurverein düsseldorf
Staff: VDI is composed of 77000 members, 1200 supporting members
Head: Dr. Ing. R. Menger
Documentation: bibliographic data
Subject(s): engineering; chemistry; environment
UDC 62; UDC 66; UDC 574
Language(s): De; En; Fr (and others)
Vocabularies: Terminology work carried out only marginally. Maintenance of the "VDI-Thesaurus Technik"
Publications: *VDI-Nachrichten.* A weekly journal concerned with technology, economics and science. 52 issues annually;

VDI-Z - Entwicklung - Konstruktion - Produktion (Journal of the Verein Deutscher Ingenieure für Maschinenbau und Metallverarbeitung). 24 issues annually; and a number of other journals and documents in the engineering field
Services: reference; referral; other: SDI, on-line service via terminal, and others
User aids: Newsletter

○ ***351*** *(Organization)*
Amt für Standardisierung, Meßwesen und Warenprüfung.
ASMW
Wallstraße 16
Berlin, 102
GDR
Head: Prof. Dr. Lilie
Vocabulary standards:
Subject(s): all fields
UDC 0/9
Language(s): De

○ ***352*** *(Organization)*
Technische Universität Dresden. Institut für Angewandte Sprachwissenschaft
[Technical University of Dresden. Institute of Applied Linguistics]
Parent organization: Ministerium für das Hoch- und Fachschulwesen der DDR [Ministry of Education for Universities and Specialized School Levels of the GDR]
Mommsenstraße 13 Phone: 4636020
Dresden, 8027 Telex: 02278
GDR
Staff:
Head: Prof. Dr. E. Baumann
Founded: 1951
Documentation:
bibliographic data
factographic data
Subject(s): chemistry; chemical technology; electrical engineering; electronics
UDC 54; UDC 66; UDC 621. 3; UDC 681. 3
Language(s): Ru; En; De
Training: undergraduate; graduate; postgraduate
Access to training: available to outside users
Curriculum: fundamentals of terminology; terminological principles; lexicographical methods. Courses on science of terminology, lexicography, standardization of terms as well as courses on the principles of terminology and specialized terminology. These courses are subjects taught at the university, both in Ru and En
Publications: *Deutsche Fachsprache der Technik - Ein Ratgeber für die Sprachpraxis* [German Specialized Vocabulary of Technology - A Guide to Practical language. Leipzig: VEB Verlag Enzyklopädie, 1975.
In addition: Several dictionaries, all published by VEB
Services: reference; loan; translation; other: Terminological advice
No service charges: (some restrictions)

○ ***353*** *(Data Bank)*
Technische Universität Dresden. Institut für Angewandte Sprachwissenschaft
[Technical University of Dresden. Institute of Applied Linguistics]
Parent organization: Ministerium für das Hoch- und Fachschulwesen der DDR [Ministry of Education for Universities and Specialized School Levels of the GDR]
Mommsenstraße 13 Phone: 4636020
Dresden, 8027 Telex: 02278
GDR
Purpose of data bank/users: Multiple purpose (translation aid, etc.); Translators, subject specialist, information specialists
Data base description:
Name: Elektronisches Wörterbuch der Fachsprachen (EWF) [Electronic Dictionary of Specialized Languages]
Classification: Special polyhierarchical classification for special fields
Subject(s): chemistry; electrical engineering; electronics; hydraulics; pneumatics
UDC 54; UDC 66; UDC 621. 3; UDC 681. 3
Language(s): Ru; En; De

● ***354*** *(Organization)*
Institut für Technologie und Warenwirtschaftslehre der Wirtschaftsuniversität Wien
[Institute for Technology and Commodity Science of Vienna University of Economics (not official)]
Parent organization: Wirtschaftsuniversität Wien; Internationale Gesellschaft für Warenkunde und Technologie
Augasse
Wien, A-1090
Austria
Staff: 10
Head: Univ. Prof. Dr. Josef Hölzl,
Expert(s): Ing. Dkfm. Dr. Helge Gasthuber
Staff categories:
1 terminologist(s) (special field); 1 terminologist(s) (theory)
Founded: 1919
Documentation: bibliographic data
Subject(s): commodiy; commodity classification; product classification
UDC 620. 2
Language(s): De; En; It
Publications: *Schriftenreihe des Institutes für Technologie und Warenwirtschaftslehre der Wirtschaftsuniversität Wien*, edited by Univ. Prof. Dr. J. Hölzl. FORUM WARE. *Internationale Zeitschrift für Warenwissen*, edited by Deutsche, Österreichische und Internationale Gesellschaft für Warenkunde und Technologie. Ulm: Verlag FORUM WARE, appears 4 times a year
Services: reference (publications concerning product terminology and classification)
No service charges
Operation:
manual

(Project(s))
Title: Beschreibung und Klassifikation von Produkten [Description and classification of products]
Project type: institutional; in progress
Principal investigator: Prof. Dr. J. Hölzl
Collaborators: 2
Scope and/or subject matter: Elaboration of a theoretical basis of an adequate description and classification of products (commodies)
Method of work: Theoretical (desk) research: analysis of literature, elaborating a hypothethical concept for product description and classification, elaborating some empirical test for validating theoretical hypotheses
Source material: Special literature in the field of commodity science
Publications: HÖLZL, J. *Name und Benennung in der Warenklassifikation*. In: *Studien zur Klassifikation*, Vol. 5, Frankfurt/Main 1979, p. 155–170.
GASTHUBER, H. *Die Begriffe Eigenschaft und Merkmal in der Warenklassifikation und -beschreibung*. In *Studien zur Klassifikation*, Vol. 10, Frankfurt/Main 1982.; Working papers and interim reports on request

○ ***355*** *(Organization)*
Österreichisches Normungsinstitut
[Austrian Standards Institute]
ON
Heinestraße 38
Postfach 130 Phone: 43 222 33 55 19
Wien, A-1021 Telex: 115 960 onorm a
Austria Telegr.: austrianorm
Head: Techn. Rat Ing. Josef Maurer, Director

Vocabulary standards:
Subject(s): all fields
UDC 0/9
Language(s): De

● *356* *(Organization)*
Jazykovedný ústav Ľudovíta Štúra
[Linguistic Institute Ľd. Štú (not official); Institut jazykoznanija imeni Ľudovita Štúra (Ru)]
JÚĽŠ SAV
Parent organization: Slovak Academy of Sciences
Nálepkova 26 Phone: 59671
Bratislava, 81364
Czechoslovakia
Head: Dr. J. Kačala, CSc.
Expert(s): Dr. I. Masár
Founded: 1943
Documentation: bibliographic data
Subject(s): all fields
UDC 0/9
Language(s): Sk
Vocabulary standards:
Subject(s): all fields
UDC 0/9
Language(s): Sk
Theory: Sk

● *357* *(Commission/Committee)*
Medzinárodná komisia pre lingvistickú terminológiu
7International Committee for Linguistics Terminology; Meždunarodnaja kommissia po lingvističeskoj terminologii (Ru)]
MTKL
Parent organization: Meždunarodnyj komitet slavistov [International Committee of Slavists]
Visiting address: Bratislava, Nálepkova 26
Chairperson: Prof. Dr. J. Horecký, DrSc.
Established: 1958
Vocabularies:
Subject(s): linguistics
UDC 80
Language(s): Ru; CS; Sk; En; Fr; De; other Slavic languages
Publications: *Slovník slovanské lingvistické terminologie* [Dictionary of Slavic linguistics terminology]. I + II. Praha: Academia, 150 p. + 150 p. *Aktuálne otázky jazykovednej terminológie* [Problems of linguistic terminology]. Bratislava: 1976; Göteborg: 1980.

● *358* *(Organization)*
Katedra Neslovanských Jazyku
[Institute for Non-Slavic Languages (not official)]
KNJ
Parent organization: Vysoká Skola Zemědělská Praže (School for Agriculture) Phone: 32 79 812
Praha 6 - Suchdol, Cs-16021
Czechoslovakia
Staff: 9
Head: Prof. Dr. Jaroslav Voráček
Expert(s): Prof. Dr. Jaroslav Voráček
Staff categories:
1 terminologist(s) (special field) ; 3 linguist(s) ; 6 terminology teacher(s)
Founded: 1953
Documentation:
bibliographic data
factographic data
Subject(s): agriculture
UDC 63
Language(s): De; En; Fr; Es
Theory: Cs; De
Training: undergraduate; postgraduate
Curriculum: fundamentals of terminology, audio-visual aids are being used (tapes, language labs)
Services: reference; loan (restricted); other: Interpretation within the scope of school
No service charges
Operation:
manual

(Project(s))
Title: Subject will be: English teaching material for agriculture students (probable title: Angličtina I, Angličtina II)
Principal investigator: Prof. Dr. Jaroslav Voráček
Collaborators: 3
Scope and/or subject matter: Teaching material for agricultural students ; agricultural texts, adapted or simplified; grammar; exercises; special exercises on tapes as well as internal texts are to be available on tape
Method of work: Scanning existing material; selection and adaption of appropriate texts; elaboration of exercises (making use of already existing internal text books)
Source material: Internal text books and other sources available

● *359* *(Commission/Committee)*
Komise odborné terminologie v zemědělství, lesním hospodářství a potravinářském průmyslu
[Commission for special terminology in agriculture, forestry and food industry; Commission pour la terminologie technique pour l'agriculture, la sylviculture et les industries alimentaires; Kommission für Fachterminologie für Landwirtschaft, Forstwirtschaft und Nahrungsmittelindustrie (De)]
Parent organization: Institute of Scientific and Technical Information for Agriculture
Slezská 7 Phone: 257541–9, 254451–9
Praha 2, 120 56
Czechoslovakia
Visiting address: Ing. Oto Šilhán, CSc., ISTIA Director, Slezská 7, 120 56 Praha 2, Czechoslovakia
Committee members: 18
Chairperson: Prof. Dr. Ing. Alois Grolig, DrSc.
Vice-chairperson: Ing. Anton Laffèrs, CSc.
Established: 1965
Vocabularies:
Subject(s): agriculture; forestry; game management; food industry
UDC 63; UDC 663/665
Language(s): Cs; Sk; plus equivalents in Ru, En, De
Vocabulary standards: Participation in the preparation of Czechoslovak terminological standards
Subject(s): agriculture; forestry; game management; food industry
UDC 63; UDC 663/665
Language(s): Cs; Sk
Publications: *Tschechoslowakisches Wörterbuch der Pferdezucht mit deutschen Äquivalenten.* Praha: ČSAZ-ÚVTIZ (1970), 208 p.; *Czech-Slovak terminological dictionary of plant pathology and plant protection. Virology. With Russian, English, French and Spanish equivalents.* Praha: ČSAZ-ÚVTIZ (1971), 311 p.; *Tschechisch-Slowakisches terminologisches Wörterbuch - Exterieur der großen landwirtschaftlichen Nutztiere einschließlich Schafe und Ziegen mit russischen und deutschen Äquivalenten.* Praha: ČSAZ-ÚVTIZ (1973), 292 p.; *Czech and Slovak terminological dictionary - Cattlebreeding. With Russian, German and English equivalents.* Praha: ČSAZ-ÚVTIZ (1976), 366 p.; *Czech and Slovak terminological dictionary - General zootechnic and genetics. With Russian, German and English equivalents.* Praha: ČSAZ- ÚVTIZ (1979), 408 p.; *Názvoslovný zpravodaj zemědělský.* Praha: ÚVTIZ (issued in Czech annually since 1967)

(Project(s))
Responsible: Commission for special terminology in agriculture, forestry and food industry
Title: Komplexní zpracování a sjednocení odborné zemědělské, lesnické a potravinářské terminologie [Elaboration and unification of special terminology in agriculture, forestry and food industry]
Project type: institutional; in progress

Principal investigator: Ing. Oto Šilhán, CSc.; Prof. Dr. Ing. Alois Grolig, DrSc.
Collaborators: 4 internal and about 700 extramural collaborators
Scope and/or subject matter: Agriculture; forestry; food industry
Method of work: Excerption of technical terms, working out indexes, defining, unifying terms where duplicity and multiplicity exists, adding foreign language equivalents to the terms, selecting terms for special field terminological dictionaries
Source material: Special literature in the respective fields

○ *360* *(Organization)*
Úřad pro normalizaci a měřeni
[Institute for Standardization and Measurements]
ÚNM
Václavské náměsti 19 — Phone: 42 2 26 22 51
Praha 1, Cs-113 47 — Telex: c 121948
Czechoslovakia — Telegr.: normalizace
Head: Mr. T. Hill, President
Vocabulary standards:
Subject(s): all fields
UDC 0/9
Language(s): Cs; Sk

● *361* *(Organization)*
Ústav pro jazyk český
[Institute of the Czech Language (not official); Institut de la langue tchèque (not official)]
ÚJČ
Parent organization: Československá Akademie věd (Czechoslovak Academy of Sciences)
Letenská 4 — Phone: 53 93 53
Praha, CS-118 51
Czechoslovakia
Staff: 98 (entire Institute)
Head: Prof. PhDr. Jan Petr
Expert(s): Dr. Miroslav Roudný
Staff categories:
2 terminologist(s) (theory and spezial field)/linguist(s); 3 terminologist(s) (theory and spezial field) / linguist(s) / information specialist(s) / terminology teacher(s) (together 5 persons, one for terminology of linguistics)
Founded: 1946 (partly 1911)
Documentation:
bibliographic data
factographic data
Subject(s): social sciences; linguistics; natural sciences; medicine; technology; industry; agriculture; schoolbooks; dictionaries; standards
UDC 0/9
Language(s): Cs; Sk; Ru; En; Fr; De
Vocabularies:
Subject(s): linguistics; mining
UDC 80; UDC 622
Language(s): Cs
Vocabulary standards:
Subject(s): all fields
UDC 0/9
Language(s): Cs; Sk; Ru; En; Fr; De; Es
Theory: Cs (linguistics in 14 languages)
Publications: *Přiruční slovnik jazyka českého*, 1935–57, 9 vol., publ. by Pedag. rekl.; *Slovnik spisovného jazyka českého*, 1958–1970, 4 vol., publ. by Academia;
Slovnik spisovné češtiny, 1978, 1 vol., publ. by Academia;
Tvořeni slov v češtiné,, 1962–67, 2 vol., publ. by Academia.
Services: reference; other: Consultations, instructions and proof-reading in terminology
Service charges: In some cases
Operation:
manual
Terminological data:
definitions; explanations; context; other data elements (grammatical gender, and other grammatical details)

● *362* *(Commission/Committee)*
Ústav pro jazyk český - komisi
Parent organization: Ústav pro jazyk cesky
Letenská 4 — Phone: 53 93 53
Praha, CS- 118 51
Czechoslovakia
Vocabularies:
Subject(s): packaging; pumps; organic and inorganic chemistry; containers; animal feeding; law; trains; agriculture; geodesy; etc.
UDC 0/9
Language(s): Cs; Sk; Ru; En; Fr; De; Es

● *363* *(Organization)*
Ústředí vědeckých, technických a ekonomických informací
[Central Office of Scientific, Technical and Economic Information; Office central de l'information scientifique, technique et economique; Zentralstelle für wissenschaftliche, technische und ökonomische Information (De)]
UVTEI
Konviktská 5 — Phone: 266351/-4
Praha, 113 57 — Telex: Praha 122214
Czechoslovakia — Telegr.: UVTEI Praha
Staff:
Head: Ing. Jaroslav Volný
Expert(s): Ing. Jiři Fogl, CSc.
Staff categories:
1 documentalist(s)/librarian(s)
Founded: 1966
Documentation:
bibliographic data
factographic data
Subject(s): information science; librarianship; terminological control; theoretical problems of terminology and lexicography
UDC 002; UDC 007; UDC 02
Language(s): Cs; Sk; Ru; En; Fr; De
Vocabularies:
Subject(s): information science; special librarianship
UDC 002; UDC 007; UDC 02
Language(s): Cs; Sk; Ru; En
Vocabulary standards:
Subject(s): information science; special librarianship
UDC 002; UDC 007; UDC 02
Language(s): Cs
Theory: Cs
Publications: Since 1978: ČSN 01 0183 - *Informačni prameny. Názvoslovi* [Czechoslovak State Standard - Information Sources. Terminology]. 1978, 55 p.; HLAVATÁ E. et al. *Informační prameny. Terminologický slovnik* [Information Sources. Vocabulary of Terminology]. Praha: ÚVTEI, 1979, 331 p.
Services: reference; loan; reproduction; referral; only secondary information from information science and special librarianship including terminology of this branch
Data base description: Terminological file with definitions of concepts from information science, special librarianship and adjacent disciplines
Name: 15 000 excerpts
Language(s): Cs; Sk
Classification: Alphabetical ordering, classification used in: Dictionary of Informatics Terms, Moscow 1975, ICSTI
Terminological data:
definitions; explanations; context; equivalents (in special cases); other data elements (bibliographic references)
Access:
via telephone / via telex
Data available:
on exchange basis (for foreign users); free of charge (for Czechoslovak users)

● *364* *(Organization)*
Výskumné výpočtové stredisko
[Computing Research Centre; Forschungsrechenzentrum (De)]

VVS; CRC
Dúbravská 3
P. O. Box 1101 Phone: 46531
Bratislava, Cs-88531 Telex: 093587 VOSN C
Czechoslovakia
Staff:
Head: Prof. Anton Klas
Expert(s): Dr. Theo Hlaváč
Staff categories:
4 translator(s); 1 terminologist(s) (special field); 1 linguist(s)
Founded: 1968
Subject(s): data processing; economics; statistics; information systems
UDC 007; UDC 31; UDC 33; UDC 681. 3
Language(s): Sk; En; De; Ru
Vocabulary standards:
Subject(s): data processing
UDC 007; UDC 31; UDC 33; UDC 681. 3
Language(s): Sk; Ru; En; De
Publications: More than 80 reports each containing a glossary providing definitions of the terms used.
Dictionary: *Počitače a spracovanie dát* [Computers and data processing]. Part I: En-Sk, Part II: Sk-En; *Slovnik terminov z národohospodárskych bilancii a národnýh účtov* (Glossary of terms used in national economy balances and national accounts). No. 104/1: En-Sk-Fr-Ru, No. 104/2: Sk-En-Fr-Ru;
From 1976: quarterly: *Terminologické listy* [Terminological sheets], terms in: Sk-En-Ru-De-Fr, definitions in: Sk
Operation:
computer-assisted
Data base description:
Language(s): Sk; En
Terminological data:
equivalents
Data processing:
Hardware: CDC; 3300
Software:
programming language(s): COBOL

● *365* *(Commission/Committee)*
Terminologická komisia VVS
[Terminological Commission CRC]
TK VVS; TC CRC
Parent organization: Výskumné výpočtové stredisko [Computing Research Centre]
Dúbravská 3
1101 Phone: 46531
Bratislava, 88531 Telex: 093587 VOSN C
Czechoslovakia
Committee members: 20
Chairperson: Dr. Theo Hlaváč
Established: 1970
Vocabularies:
Subject(s): data processing; economics; statistics; informations systems
UDC 007; UDC 31; UDC 33; UDC 681. 3
Language(s): Sk; En; De; Ru
Vocabulary standards:
Subject(s): data processing
UDC 681. 3
Language(s): Sk; En; De; Ru
Publications: *Terminologické listy* [Terminlogical sheets], quarterly; without charge

● *366* *(Organization)*
Instytut Informacji Naukowej, Technicznej i Ekonomicznej
7Institute for Scientific, Technical and Economic Information]
IINTE; ISTEI
Parent organization: Centrum Informacji Naukowej, Technicznej i Ekonomicznej
Al. Niepodległości 188
Warszawa, 00–950
Poland
Staff: 90 (entire institute)
Head: Prof. Dr. Konrad Fiałkowski
Expert(s): Krystyna Tittenbrun, M. A.
Staff categories:
1 terminologist(s) (theory); 66 documentalist(s) / librarian(s) / linguist(s) / information specialist(s)
Founded: 1972 (Originally founded in 1950 as Main Institute of Scientific and Technical Documentation)
Documentation:
bibliographic data
factographic data
Subject(s): information science (and related fields)
UDC 002; UDC 007
Language(s): Pl; En; Fr; De; Ru
Vocabularies:
Subject(s): information science
UDC 002; UDC 007
Language(s): Pl (with equivalents in En, Ru, De and Fr)
Vocabulary standards:
Subject(s): information science
UDC 002; UDC 007
Language(s): Pl
Publications: *Słownik terminologiczny informacji naukowej.* [Dictionary of information science terms]. Warszawa: Instytut Informacji Naukowej, Technicznej i Ekonomicznej, 1979, 237 p.; Tittenbrun, K. *Podstawowe źródła leksykograficzne z zakresu informacji naukowej. Monografia.* [Basic lexicographical sources of scientific information. A monography]. Warszawa: Instytut Informacji Naukowej, Technicnej i Economicznej, 1979. Prace IINTE 24 [ISTEI Reports 24]; Tittenbrun, K. *Problemy terminologiczne informacji naukowej na tle ogólnych zagadnień słownictwa specjalnego.* [Terminology problems of information science against a background of special vocabulary general questions]. In: *Informacja naukowa - wybrane prace badawcze Instytutu IINTE za lata 1978–79* [Information science - selected ISTEI research reports for 1978–79]. Warszawa: IINTE, 1980, p. 29–42. Prace IINTE 34 [ISTEI Reports 34]; Tittenbrun, K. *Polskie piśmiennictwo terminologiczne z zakresu informacji naukowej i dziedzin pokrewnych. Bibliografia adnotowana 1945 - 1978.* [Polish terminological literature in the area of information science and related fields. An annotated bibliography 1945 - 1978]. ISTEI Reports, in print.
Theory: Pl
Services: reference
No service charges
Terminological data:
definitions; context; equivalents
Data processing:
Hardware: RIAD 32; IBM 360/50; mini MERA 400; 32 (RIAD); 360/50 (IBM); 400 (mini MERA)
Software: ISIS; own software
Access:
via telephone

● *367* *(Commission/Committee)*
Polska Akademia Nauk. Komitet terminologii
[Polish Academy of Sciences. Committee on Terminology]
KT PAN
Parent organization: Polska Akademia Nauk
Nowy Świat 72 Phone: 26 52 31 ext. 141
Warszawa, 00330
Poland
Contact: Prof. Dr. B. Walentynowicz
Institute of Philosophy and Sociology
apt. 304
00330 Warszawa
Committee members: 60 experts from all fields of science, technology and practice
Chairperson: Prof. Dr. Janusz Groszkowski
Vice-chairperson: Prof. Dr. Jerzy Skowronski; Prof. Dr. Wacław Szymczak; Mr. Bolesław Adamski
Established: 1979
Vocabularies: Yes
Vocabulary standards:
Subject(s): science; technology; engineering
UDC 5/7

Language(s): Pl
Theory: Yes
Publications: In preparation

○ ***368*** *(Organization)*
Polski Komitet Normalizacij i Miar
[Polish Committee for Standardization, Measures and Quality Control]
PKNiM
Ul. Elektoralna 2 Phone: 48 22 20 54 34
Warszawa, PL-00–139 Telex: 81 36 42 pkn pl
Poland Telegr.: pekanim
Head: Mr. Franciszek Szlachcic, President
Vocabulary standards:
Subject(s): all fields
UDC 0/9
Language(s): Pl

● ***369*** *(Commission/Committee)*
Polski Komitet Terminologii Elektrycznej
[Polish Committee on Electrical Terminology]
PKTE
Parent organization: Stowarzyszenie Elektryków Polskich [Association of Polish Electrical Engineers]
Czackiego 3/5
903 Phone: 26 86 02
Warszawa, 00950 Telex: 813 225 NOT PL 841
Poland Prof. Dr. B. Walentynowicz, 03936 Warszawa, Zwyciezców 3/5
Committee members: 40 experts in Electrical Engineering and Technology
Chairperson: Prof. Dr. Bohdan Walentynowicz
Established: 1917 (till 1972 Central Commission of Electrical Terminology)
Vocabularies:
Subject(s): electrical engineering; electronics
UDC 621. 3; UDC 621. 38
Language(s): Pl; En
Vocabulary standards:
Subject(s): electrical engineering; electronics
UDC 621. 3; UDC 621. 38
Language(s): Pl; En
Theory: Yes; Pl; En
Publications: *Terminologia Elektriczna* [Electrical Terminology], Bulletin - published 4 times a year, as appendix to monthly *Przeglad Elektrotechniczny* [Electrical Revue], first issue 1972 No. 1

◐ ***370*** *(Organization)*
Uniwersytet im. Adama Mickiewicza. Instytut Fizyki
[Adam Mickiewicz University. Institute of Physics]
UAM
ul. Grunwaldzka 6 Phone: 699 181, ext. 392 ; 619 18
Poznań, 60–780 Telex: 041 5260
Poland
Staff: 230
Head: Prof. Dr. habilit. Marian Surma, Director of the Institute of Physics
Expert(s): Prof. Dr. habilit. Henryk Cofta
Staff categories:
terminologist(s) (theory); translator(s)
Founded: 1919/1969
Vocabularies: physics
UDC 53
Language(s): Pl; En; Fr; De; Ru

● ***371*** *(Organization)*
Wydawnictwa Naukowo - Techniczne
[Scientific and Technical Publishers; Naučno-tehničeskoe izdatelstvo (Ru)]
WNT
Mazowiecka 2/4
P. O. Box 359 Phone: 26–72–71
Warszawa, 00–950 Telegr.: Warszawa ENTE
Poland
Staff: 23 (Technical Dictionary Division only)
Head: Mr. C. Kulesza
Expert(s): Mr. S. Czerni (Head of Dictionary Division)
Staff categories:
2 terminologist(s) (special field); 13 linguist(s); 1 terminology teacher(s)
Founded: 1949
Documentation:
bibliographic data
factographic data
Subject(s): natural sciences; technology
UDC 5/6
Language(s): Pl; Ru; En; De; Fr
Vocabularies:
Subject(s): science; technology
UDC 5/6
Language(s): PL; Ru; En; De; Fr
Theory: Pl
Publications: Scientific and technical dictionaries (about 170 titles). See *Bibliography of Dictionaries published in Bulgarian People's Republic, Czechoslovak Socialist Republic, German Democratic Republic, Hungarian People's Republic, Polish People's Republic, Romanian Socialist Republic, Union of Soviet Socialist Republics* (1945–1978)

○ ***372*** *(Organization)*
Magyar Szabványügyi Hivatal
[Hungarian Standards Institute]
MSZH
Postafiok 24 Phone: 36 1 18 30 11
Budapest 9, H-1450 Telex: 22 5723
Hungary Telegr.: normhungaria budapest
Head: Dr. Joseph Olajos, President
Vocabulary standards:
Subject(s): all fields
UDC 0/9
Language(s): Hu

◐ ***373*** *(Organization)*
Országos tervhivatal számitástechniaki központja
[Computing Centre of the Hungarian Planning Office]
OTSZK
Angol u. 27
Budapest, H-1149
Hungary
Staff: 5 (2 full time, 3 part time)
Head: Mr. Imre Tóth
Expert(s): Dr. Dénes Vargha
Staff categories:
2 mathematicians; 3 economists
Founded: 1968
Documentation: bibliographic data
Subject(s): economics
UDC 33
Language(s): Hu
Vocabularies:
Subject(s): economics
UDC 33
Language(s): Hu
Publications: VARGA, D. *Információs tezauruszok készitésének módszertana* [Methodology for preparing information thesauri]. Budapest: OMKDK, 1969, 156 p. Russian translation in: *Sbornik perevodov po voprosam informacii, teorii i praktiki*, No. 17, Moskva, VINITI; 1970, 108 p.;
VARGA, D. (ed.) *A dokumentáció nyelvészeti kérdései* [Linguistic problems of documentation]. Budapest: OMKDK, 1966, 156 p.;
VARGA, D. (ed.) *Dokumentáció és nyelvészet - az értelmi összefüggések formalizálása* [Documentation and linguistics - formalization of semantic interconnections]. Budapest: OMKDK, 1969
Services: reproduction; other: planning services
Service charges
Data processing:
Software: Standard ICL software and special software for planning and modelling purposes

● *374* *(Organization)*
Association française de normalisation
[French Association for Standardization (not official); France]
AFNOR
Tour Europe
Cedex 7 Phone: (1) 778 13 26
Paris La Défense, F-92080 Telex: 611 974F
Staff: 410
Head: M. Bernard Vaucelle
Expert(s): Mlle. Jocelyne Laurent
Staff categories:
1 terminologist(s) (special field); 2 documentalist(s)/librarian(s); 14 information specialist(s); 3 translator(s) full time (together with interpretation); 20 translator(s) free-lance part time; engineers, technicians, commercial and administrative personnel; secretaries
Founded: 1926
Documentation:
bibliographic data
factographic data
Subject(s): all fields
UDC 0/9
Language(s): Fr; En (Some terminological standards have equivalents in other languages, such as De; Ru; Es; It; etc.)
Vocabulary standards:
Subject(s): all fields
UDC 0/9
Language(s): Fr mainly; En with ISO
Theory: Fr
Publications: *Catalogue des normes françaises* (published annually); French standards (nearly 11 000); Handbooks of French standards on specific technical fields and other technical books on standardization ;
Index NORMATERM - Index permuté par mots - vedettes des termes techniques normalisés - 2 vol. - Paris: AFNOR, 1977 (updated in 1980); Liste bibliographique no. 1 *Normes de vocabulaire* - 1979 (updated in September 1980)
Services: reference; translation; other: sale of standards
Service charges
Operation:
computerized
Data base description:
Name: NORMATERM
Size: 22 000 concepts
Language(s): Fr; En
Classification: ISONET thesaurus; classification of standards and documents on standardization
Terminological data:
definitions; explanations; classified layout; standardized terms; equivalents; other data elements
Data processing:
Hardware: CII - HB; IRIS 45; 128 Kb; 200 Mb; 1600 Bpi - 9 tracks; printer; With French Network TRANSPACK for NORIANE; with AFNOR's computer for NORMATERM
Software: Used for two data bases - NORIANE and NORMATERM; operating system: CII - HB SIRIS 3; operation mode: real time; batch; demand; programming language(s): COBOL; data base management system/retrieval documentation system: MISTRAL
Access:
via telephone / via telex
Data available:
with charge

● *375* *(Data Bank)*
Normalisation - Automatisation de la terminologie
[Standardization - Automation of Terminology (not official)]
NORMATERM
Parent organization: Association Française de Normalisation (AFNOR)
Tour Europe
Cedex 7 Phone: (1) 778 13 26
Paris La Défense, F-92080 Telex: 611 974F
France
Staff: 2 especially for NORMATERM
Head: M. Bernard Vaucelle
Expert(s): Mlle Jocelyne Laurent
Staff categories:
2 documentalist(s)/librarian(s) (part time)
Founded: 1973
Purpose of data bank/users: To make available to any user the terminology standardized in AFNOR and ISO standards; Subject specialists, translators, engineers, information specialists, etc.
Operation:
computerized
Data base description:
Name: NORMATERM
Size: 23 000 concepts (i.e. 90 000 entries for both languages)
Language(s): Fr; En
Classification: ISO thesaurus; classification of standards and documents on standardization
Subject(s): all fields
UDC 0/9
Language(s): En; Fr
Terminological data:
definitions (in Fr); explanations (in Fr); classified layout (generic and specific terms, in Fr); standardized terms; equivalents (terms representing the concepts, in Fr and En); other data elements
Products: Glossaries of standardized terms; microforms (on request); magnetic tapes, disks (on request)
Services: Answers to specific queries
Service charges: For glossaries; for specific queries the charge is 150 FF for ten concepts with definition(s)
Access:
via telephone / via telex
Data available:
with charge
Data processing:
Hardware/Capacity: CII - HB IRIS 45; 128 Kb; 200 Mb; 1600 Bpi - 9 tracks; Printer; With AFNOR's computer
Software:
operating system: CII - HB SIRIS 3; operation mode: batch or real time; programming language(s): COBOL; data base management system/retrieval documentation system: MISTRAL/MISTRAL

● *376* *(Organization)*
Centre de préparation documentaire à la traduction
CPDT
16 rue Beaurepaire Phone: (1) 208. 86. 32
Paris, F-75010 Telex: 610465 Matim/
France CPDT
Staff: 10 (advisors)
Head: Mr. Gérard Pierson
Expert(s): Mr. Gérard Pierson
Staff categories:
3 terminologist(s) (theory); 1 documentalist(s)/librarian(s); 1 information specialist(s); 10 translator(s)
Founded: 1973
Documentation: factographic data
Subject(s): physics; chemistry; biology; technology; applied science
UDC 53; UDC 54; UDC 57; UDC 58; UDC 62; UDC 63; UDC 66; UDC 67; UDC 68; UDC 69
Language(s): Ch; En; Fr
Vocabularies:
Subject(s): all fields
UDC 53; UDC 54; UDC 57; UDC 58; UDC 59; UDC 60; UDC 62; UDC 63; UDC 66; UDC 67; UDC 68; UDC 69
Language(s): En; Ch; Fr; De; Ru
Publications: Periodicals: *Bulletin d'Alerte Terminologique/Current Terminology Awareness Bulletin*; *Langue Chinoise/Chinese Language*. Some of these have appeared as monographs as well.
Other publications: *Flora of the Chinese Characters* ; *Information & Translation* - IT
Colletion des Etudes Documentaires Appliquées à la Traduction Automatique et Humaine (CEDATAH); IT - *Information & Translation; Chinese Reading Aid Series*

(CRAS); *Glossaires de l'Ingénieur Traducteurs* (GIT); *Words in Context* (WIC)
Services: reference; translation; referral; SDI; other: typing: Arabic, Chinese (traditional and simplified), Cyrillic, Greek, Hebrew
Service charges: Membership fees
User aids: *Current Terminology Awareness Bulletin; Langue Chinoise/Chinese Language;* CEDATAH (Newsletter)
Operation:
manual
Data base description:
Name: CEDATAHTERM (Collection des Etudes Documentaires Appliquées à la Traduction Automatique et Humaine)
Size: About 4000
Language(s): Ch; En; Fr
Classification: Alphabetical system; Third International Classification of Patents
Terminological data:
context; equivalents; other data elements (context with reference)
Access:
via telephone / via telex
Data available:
with charge (membership fees)

(Project(s))
Title: La néologie anglo-américaine contemporaine et l'édition 1980 du Harrap's - un constat [Contemporarvy Anglo-American Neology and the 1980 edition of the English-French Harrap's - an assessment]
Project type: institutional; in progress
Principal investigator: G. Pierson
Collaborators: About 5
Scope and/or subject matter: Comparative lexicography; neology
Method of work: Using own lexicological records; comparisons with the entries of the 1980 edition of the English-French Harrap's are being carried out
Source material: Own records; Webster suppl; and other relevant sources (Barnhart, etc.)
Publications: Results are to be published in *Bulletin d'Alerte Terminologique/Current Terminology Awareness Bulletin*

● ***377*** *(Data Bank)*
Centre de préparation documentaire à la traduction
CPDT
16 rue Beaurepaire — Phone: (1)208. 86. 32
Paris, F-75010 — Telex: 610465 MATIM/CPDT
France
Staff: 10 advisors
Head: Mr. Gérard Pierson
Expert(s): Mr. Gérard Pierson
Staff categories:
3 terminologist(s) (theory); 1 documentalist(s)/librarian(s); 1 information specialist(s); 10 translator(s)
Founded: 1973
Purpose of data bank/users: Updating of dictionaries for translation purposes; Translators, subject specialists
Data base description:
Name: CEDATAHTERM
Size: appr. 4000 entries
Language(s): Alphabetical order; International Classification of Patents
Subject(s): physics; biology; technology; applied sciences
UDC 5/6
Language(s): Ch; En; Fr
Terminological data:
explanations (by context); context (with full reference); equivalents; other data elements (context with reference)
Products: Collection des Etudes Documentaires Appliquées à la Traduction Automatique et Humaine (CEDATAH); IT Information & Translation, Chinese Reading Aid Series (CRAS); Glossaires de l'Ingénieur Traducteur (4 IT); Words in Context (WIC)
microforms ("Current Terminology Awareness Bulletin/Bulletin d'Alerte Terminologique" combined with "Langue Chinoise/Chinese Language")
Services: SDI; information and referral services

● ***378*** *(Organization)*
Comité d'études des termes techniques français
[Studies committee of French technical terms (not official)]
CETTF
11, Avenue du Général Pershing
Versailles, F-78000 — Phone: 954 85 62
France
Staff: 36 full time, 32 part time
Head: Mr. Pierre Agron
Expert(s): Mr. Pierre Agron
Staff categories:
9 linguist(s); 3 translator(s); 1 information specialist (s); 20 other professional(s)
Founded: 1954
Documentation:
bibliographic data
factographic data
Subject(s): technology
UDC 6
Language(s): Fr
Vocabularies:
Subject(s): technology
UDC 6
Language(s): Fr
Publications: *Termes techniques français: essai d'orientation de la terminologie.* Paris: Editeur Hermann, 1972; 16°e, 17°e, 18°e, 19°e et 20°e séries de fiches techniques (suite de livre ci-dessus);
Le langage des techniques; *Plaidoyer pour les mots-ancêtres, Présentation du Comité d'étude des termes techniques français*; *Pour des pensers nouveaux*; *Les marques déposées et le langage*;
Humbles mots, mots de prestige, et mots de soufre; *Vite ou bien*; *Le grain des mots*;
C'est un métier que fabriquer des mots; *Entendre les mots.*
Services: reference
Service charges

● ***379*** *(Organization)*
Laboratoire d'information et de documentation en géographie 'INTERGÉO'
INTERGÉO
Parent organization: Centre National de la Recherche Scientifique (C.N.R.S.)
191, rue Saint-Jacques — Phone: (1) 633. 74. 31
Paris, F-75005
France
Staff: 18
Head: M. Michel Drain, Maître de recherche au C. N. R. S.
Staff categories:
10 terminologist(s) (theory) / documentalist(s) / librarian(s); 2 translator(s)
Founded: 1945
Documentation: bibliographic data
Subject(s): geography
UDC 91
Language(s): Fr
Vocabularies:
Subject(s): geography
UDC 91
Language(s): Fr
Vocabulary standards:
Subject(s): geography
UDC 91
Language(s): Fr
Publications: *Bibliographie géographique internationale.* Fascicules trimestriels + index récapitulatif annuel (The volumes published since 1977 are available at INTERGÉO);
Guide d'interrogation de la base bibliographique *FRANCIS.* Vol. 15 - *Bibliographie Géographique Internationale*: fasc. 1 - Méthologie; fasc. 2 - Mots-clés;

Repértoire des géographes francophones. Vol. I: FRANCE: 2nd ed. (published in January 1981);
Vocabulaire géomorphologique russe-français;
INTERGÉO-Bulletin and Lettre d'INTERGÉO (published quarterly);
Documents INTERGÉO: *La documentation cartographique en France*; *Enquête sur les périodiques français géographiques et d'intérêt géographique*;
Terminologie de l'étude des paysages (G.Haase); *L'action de l'homme sur le paysage naturel* (A.G.Isacenko); *La structure du commerce extérieur tunisien* (M.Zghal);
Le développement industriel du Sahel tunisien (M.Jedidi); *Le développement du tourisme au Sahel tunisien* (M.Jedidi);
THESAURUS: *Géographie urbaine* (1972); *Géomorphologie* (1973); *Géographie de la population* (1974); *Hydrologie* (1975);
Géographie du commerce (1976); *Géographie industrielle* (1976); *Noms de lieux* (1980); *Géographie du tourisme* (1981)
Thésaurus *Noms de lieux* [Thesaurus *Names of Places*]. This thesaurus contains about 5000 toponymes of the whole world (regions, countries, states, provinces, rivers, towns, mountains, etc.)
Services: reference; reproduction; translation (Russian); SDI
Service charges
User aids: INTERGÉO-Bulletin and Lettre d'INTERGÉO (published quarterly)
Operation:
computerized
Data base description:
Classification: Thesaurus
Terminological data:
definitions (rarely); classified layout; standardized terms; equivalents (rarely)
Data processing:
Software: operation mode: batch; real time; programming language(s): PL1
Data available:
with charge

(Project(s))
Title: Thésaurus de géographie humaine (franco-anglais) [Thesaurus for social, economic and human geography (En-Fr)]
Principal investigator: Ms. A. Voionmaa
Collaborators: 4 or 5
Scope and/or subject matter: Social, economic and human geography, about 1500 terms
Method of work: Team work
Source material: A comprehensive, entirely revised and standardized edition of previous terminological publications issued by INTERGÉO on the subjects (cf.Supra)

○ ***380*** *(Organization)*
Ente Nationale Italiano di Unificazione
7National Italian Institute of Standardization]
UNI
Piazza Armando Diaz 2 — Phone: 39 2 87 69 14
Milano, I-20123 — Telex: 312481 uni
Italy — Telegr.: unificazione
Head: Dr. Ing. Franco Beccalli, Vice-président délégué
Vocabulary standards:
Subject(s): all fields
UDC 0/9
Language(s): It

● ***381*** *(Organization)*
Instituto nacional de racionalización y normalización
[National Institute for Rationalization and Standardization]
IRANOR
Parent organization: Consejo Superior de Investigaciones Científicas
Zurbano, 46 — Phone: (1) 4104676
Madrid - 10 — Telex: 46545 UNOR E
Spain — Telegr.: IRANOR
Staff: 100
Head: Dr. Juan de la Ynfiesta Molero
Expert(s): Mr. Carlos Cerquella
Staff categories:
2 terminologist(s) (special field); 1 terminologist(s) (theory); 1 documentalist(s)/librarian(s); 1 linguist (s); 3 information specialist(s); 4 translator(s)
Founded: 1946
Documentation: factographic data
Subject(s): all fields
UDC 0/9
Language(s): Es
Vocabulary standards:
Subject(s): all fields
UDC 0/9
Language(s): Es
Publications: National standards (UNE); Catalogue UNE; *Boletín de la Normalización Española* (Spanish Standards Bulletin)
Services: reference
No service charges
User aids: Newsletter; Catalogue
Operation:
manual
computer-assisted
Data base description:
Language(s): Es
Classification: Thesaurus ISO; UDC
Terminological data:
definitions; standardized terms; equivalents
Access:
via telephone / via telex
Data available:
on exchange basis; free of charge

● ***382*** *(Commission/Committee)*
Terminología
[Terminology]
IRN/CT 1
Parent organization: Instituto Nacional de Racionalización y Normalización (IRANOR)
Zurbano, 46 — Phone: (1) 4104676
Madrid-10 — Telex: 46545 UNOR E
Spain — Telegr.: IRANOR
Contact: Mr. L. Montero
Committee members: 5
Chairperson: Mr. Carlos Cerquella
Established: 1975
Vocabulary standards:
Subject(s): all fields
UDC 0/9
Language(s): Es
Publications: Draft Standards UNE

● ***383*** *(Organization)*
Junta de energía nuclear, Ministerio de industria y energía
JEN
Av. Complutense, s/n — Phone: 2441200
Madrid 3 — Telex: 23555 JUVIG E
Spain — Telegr.: JENSEDE
Head: Dr. Manuel López Rodríquez, General Manager
Expert(s): Dr. Manuel López Rodríquez
Staff categories:
8 terminologist(s) (special field); 8 terminologist(s) (theory); 9 documentalist(s)/librarian(s); 1 information specialist(s)
Founded: 1951
Documentation:
bibliographic data
factographic data
Subject(s): nuclear energy (peaceful use)
UDC 621. 039
Language(s): Es
Vocabularies:
Subject(s): nuclear energy (peaceful use)
UDC 621. 039
Language(s): Es

Vocabulary standards:
Subject(s): nuclear energy (peaceful use)
UDC 621. 039
Language(s): Es
Theory: Es
Publications: *Lexicología Nuclear*, 1974, ISBN 84–500–6198–9; *Léxico de Términos Nucleares*, 1974, ISBN 84–500–6295–0; *Diccionario Nuclear*, 1979, ISBN 84–500–3077–3.

(Project(s))
Responsible: Comision asesora de terminología nuclear
Title: Terminología nuclear
Project type: commissioned project: (Junta de Energia Nuclear); in progress

● *384* *(Commission/Committee)*
Comisión asesora de terminologia nuclear
CATN
Parent organization: Junta de Energía Nuclear
Aven. Complutense — Phone: 2441200
Madrid 3 — Telex: 23555 JUVIG E
Spain — Telegr.: JENSEDE
Committee members: 8
Chairperson: Dr. Manuel López Rodríquez
Established: Comisión de Vocabulario: 1964; CATN 1976
Vocabularies:
Subject(s): nuclear energy (peaceful use)
UDC 621. 039
Language(s): Es
Vocabulary standards:
Subject(s): nuclear energy (peaceful use)
UDC 621. 039
Language(s): Es
Theory: Es
Publications: *Lexicologia Nuclear*, 1974, ISBN 84–500–6198–9; *Léxico de Términos Nucleares*, 1974, ISBN 84–500–6295–0; *Diccionario Nuclear*, 1979, ISBN 84–500–3077–3.

○ *385* *(Organization)*
Direcção-Geral da qualidade
[Office of Standardization]
DGQ
Rua José Estêvão, 83–A — Phone: 351 19 53 98 91
Lisboa Codex, P-1199 — Telex: 13042 qualit p
Portugal
Head: M. A. Santos Gonçalves
Vocabulary standards:
Subject(s): all fields
UDC 0/9
Language(s): Pt

○ *386* *(Organization)*
Gosudarstvennyj Komitet Standartov SSSR
[USSR State Committee for Standards]
GOST
Leninsky Prospekt 9 — Phone: 236 40 44
Moskva, SU-117049 — Telex: 411 378 gost su
USSR — Telegr.: moskva standart
Head: Prof. Dr. techn. sc. V. V. Boitsov, President
Vocabulary standards:
Subject(s): all fields
UDC 0/9
Language(s): Ru

● *387* *(Organization)*
Kafedra anglijskogo jazyka
[Department of English]
MGU im. M. V. Lomonosova
Parent organization: Filologičeskij fakul'tet Moskovskogo gosudarstvennogo universiteta (Moscow State University, Philological Faculty)
Leninskie Gory
P. O. Box B-234 — Phone: 139–20–36
Moskva, 117234
USSR
Staff: The work is done by the language-teachers with the assistance of scientists whenever necessary
Head: Prof. Olga Akhmanova, Head of Dept.
Expert(s): Dr. Sergey Grin'ov
Founded: 1966
Documentation:
bibliographic data
factographic data
Subject(s): linguistics; philology; mathematics; physics; biology; chemistry; geography; building; philosophy
UDC 1/9
Language(s): Ru; En; Fr; De; Es
Vocabularies:
Subject(s): linguistics; philology; mathematics; physics; biology; chemistry; geography; building; philosophy
UDC 1/9
Language(s): Ru
Theory: En; Ru
Curriculum: fundamentals of terminology; terminological principles; lexicographical methods (fundamentals of terminology course "Fundamentals of scientific and technical terminology" on the undergraduate level including terminological principles, methods of terminological lexicography is held at the Chair of General and Comparative Linguistics by Dr.S.V.Grin'ov). Teaching material available).
Publications: No publications since 1978
Operation:
manual
Data available:
on exchange basis

● *388* *(Organization)*
Vsesojuznyj centr perevodov naučno-tehničeskoj literatury i dokumentacii Gosudarstvennogo komiteta SSSR po nauke i tehnike i Akademii nauk SSSR
[All-Union Institute for Translation of Scientific and Technical Literature and Documentation of the State Committee of the USSR; Allunionszentrum für Übersetzungen wissenschaftlich-technischer Literatur und Dokumentation (De)]
VCP; WZP
Parent organization: The State Committee of the USSR for Science and Technology and the Academy of Sciences of the USSR
14, korp. 1, Krzhizhanovskogo
P. O. Box 117218 — Phone: 124–70–24
Moskva
USSR
Head: Prof. Yuri N. Marchuk
Expert(s): Dr. Ivan I. Oubine
Staff categories:
terminology teacher(s) (undetermined)
Founded: 1974
Documentation: factographic data
Subject(s): science; technology
UDC 5/6
Language(s): Ru; En; Fr; De; Ja; (and many other languages)
Vocabularies:
Subject(s): all fields
UDC 0/9
Language(s): Ru; En; Fr; De; Ja; (all in all about 30 languages)
Theory: Ru
Publications: Publications planned for 1983:
Bibliografičeskij ukazatel' Naučno - tehničeskie perevody 12 vypuskov [Bibliographical index Scientific and technical translations, 12 volumes].
Anglo-russkie terminy po provodnoj i dal'nej svjazi [English-Russian terms of telecommunications)
Anglo-russkie terminy po svarke [English - Russian terms of welding];
Anglo-russkie terminy po tehnologii proizvodstva radioelek-

tronnoj apparatury [English-Russian terms of the production of radio-electronic devices];
Anglo-russkie terminy po mehanike razrušenija [English-Russian terms of destructive mechanics];
Anglo-russkie terminy po ohrone vod [English-Russian terms for water protection]; *Anglo-russkie terminy po gidroskopičeskim priboram* [English-Russian terms of hydroscopic devices];
Anglo-russkie terminy po programmnomu upravleniju metallorežuščim oborudovaniem [English-Russian terms of automatic control of metal equipment];
Anglo-russkie terminy po akustike [English-Russian acoustics terms];
Nemecko-russkie terminy po organizacii i bezopasnosti dorožnogo dviženija [German-Russian terms of organization and traffic safet];
Russko-francuzskie terminy po remontno-ekspluatacionnoj dokumentacii i tehnike bezopasnosti [Russian-French terms of the establishment and use of safety documentation and technology];
Japonsko-russkie terminy po elektrohimii i korrozii [Japanese-Russian terms of electrochemistry and corrosion];
Osnovnye problemy naučno-tehničeskogo perevoda/obzor [Main problems of scientific and technical translation / survey];
Problemy terminologičeskoj leksikografii/obzor [Problems of terminological lexicography/survey]
Services: reference; reproduction; translation

(Project(s))
Title: MULTILEKS [Multilex]
Project type: commissioned project (State Committee of the USSR for Science and Technology); in progress
Principal investigator: Dr. Ivan I. Oubine
Scope and/or subject matter: Multilanguage computerized dictionaries for various subject fields
Source material: Specialized dictionaries, terminological word lists, terms collected by translators and revisors of VCP

● ***389*** *(Data Bank)*
Laboratorija leksikografii
[Lexicography Laboratory; Laboratoire lexicographique]
Parent organization: Vsesojuznyj centr perevodov naučno-tehničeskoj literatury i dokumentacii (All-Union Centre for Translation of Scientific and Technical Literature and Documentation)
14, korp. 1, Krzhizhanovskogo
P. O. Box 117218 Phone: 124–70–24
Moskva
USSR
Expert(s): Dr. Ivan I. Oubine
Purpose of data bank/users: Production of computerized dictionaries; Translators, revisors, subject specialists
Operation: computer-assisted
Data base description:
Name: MULTILEX
Size: 120. 000 entries
Language(s): Ru; En; Fr; De
Subject(s): computer technology; programming languages
UDC 681. 3
Language(s): Ru; En; Fr; De
Terminological data:
definitions; explanations; context; equivalents; other data elements
Products: Glossaries and word lists; magnetic tapes, disks
Access:
off-line
Data available:
on exchange basis
Data processing:
Hardware/Capacity: ES 1033
Software: programming language(s): PL/1

○ ***390*** *(Organization)*
Vsesojuznyj Naučno-issledovatel'skij Institut Tehničeskoj Informacii, Klassifikacii i Kodirovanija
[All Union Research Institute for Engineering Information, Classification and Coding]

VNIIKI
Parent organization: Gosudarstvennyj Komitet Standartov Soveta Ministrov S. S. S. R.
Ul. Ščuseva 4
Moskva
USSR
Head: Prof. Dr. V. Nesterov
Expert(s): Mr. V. I. Fedosimov; Ms. N. I. Volkova
Founded: 1964
Documentation:
bibliographic data
factographic data
Subject(s): all fields
UDC 0/9
Language(s): Ru; De; En; Fr (planned: Ro; Bg; Cs; Pl; Sk)
Vocabularies:
Subject(s): all fields
UDC 0/9
Language(s): Ru; De; En; Fr (planned: Ro; Bg; Cs; Pl; Sk)
Vocabulary standards: Numerous vocabulary standards in preparation
Subject(s): all fields
UDC 0/9
Language(s): Ru
Publications: BELAHOV, L. JU; POPOV-ČERKASOV, I. N. *Naučno-tehničeskaja terminologija. Sbornik standardizo vannyh i rekomenduemyh terminov v 10 tomah* [Scientific and engineering terminology. Collection of standardized and recommended terms in 10 volumes]. Moskva: Izd. Standartov, 1970/71;
BELAHOV, L. JU; POPOV-ČERKASOV, I. N. *Problemy gosudarstvennoj standardizacii terminologii v SSSR* [Problems of standardization of terminology by the state in the USSR]. Moskva: Izd. Komiteta Standartov Mer i Izmeritel'nyh Priborov pri Sovete Ministrov SSSR, 1968, 196 p.;
BELAHOV, L. JU. *Standartizacija terminologii v SSSR i za rubežom. Obzor* [Standardization of terminology in the USSR and abroad. A Synopsis]. Moskva: Izd-vo Komiteta standartov, Mer i Izmeritel'nyh Priborov pri Sovete Ministrov SSSR, 1966, 66 p.;
GOSSTANDART SSSR. VNIIKI. *Avtomatizirovannaja sistema informacionno-terminologičeskogo obsluživanija* (SBT ASITO) [Terminology data bank of the computerized system of the terminology information service (SBT ASITO)]. Naučno-tehničeskaja terminologia, 1975, n 8 (65), p. 3–12;
GOST. VNIIKI. *Metodičeskie rekomendacii po terminologičeskomy kontrolju normativo-tehničeskih documentov* [Methodological recommendations for terminological examination of technical and standard documents]. Moskova: GOST. VNIIKI, 1973, 21 p.;
VNIIKI. *Informacionno-poiskovaja sistema na mašinnom slovare standartizovannoj i rekomenduemoj naučno-tehničeskoj terminologii* [Computer retrieval system for vocabulary of standardized and recommended scientific and technical terminology]. Naučno-tehničeskaja terminologia 1 (45), 1974, pp. 41–43;
VNIIKI. *Naučno-tehničeskaja terminologija. Perečen' inostrannyh standartov* [Scientific and engineering terminology. List of foreign standards]. Moskva: Izd. Standartov, 1968, 92 p.;
VNIIKI. *Naučno-tehničeskaja terminologija. Perečen' inostrannyh standartov. Vypusk 2* [Scientific and technical terminology. List of foreign standards. Volume 2]. Moskva: Izd. Standartov, 1970, 171 p.;
VNIIKI. *Standartizacija naučno-tehničeskoj terminologii* [Standardization of scientific and technical terminology]. Moskva: Izd. Komiteta Standartov Mer i Izmeritel'nyh Priborov pri Sovete Ministrov SSSR, 1970, 220 p.;
Periodicals: *Informacionnyj ukazatel' gosudarstevennyh standartov SSSR* [Information list of state standards of the USSR]. Moskva: Izd. Standartov, appears monthly; *Naučno-tehničeskaja terminologija* [Scientific and technical terminology]. Moskva: Izd. Standartov, appears monthly.
Services: reference; other: SDI services; advice on the elaboration of vocabularies

○ *391* *(Data Bank)*

Vsesojuznyj Naučno-issledovatel'skij Institut Tehničeskoj Informacii, Klassifikacii i Kodirovanija (VNIIKI)
Ul. Ščuseva 4
Moskva K-1
USSR
Head: Mr. V. I. Fedosimov
Purpose of data bank/users: Accumulation and systematization of standardized terminology and elaboration of an automatized system for terminological service designed for organizations and enterprises of the national economy. Storage of all vocabulary standards published in USSR, CMEA and ISO vocabulary standards.; Authors of technological, normative and other documents, translators, editorial workers, authors of text-books and manuals; Terminological data bank serves more than 150 collective subcribers
Operation:
computerized
Data base description:
Name: Avtomatizirovannaja sistema informacionno-terminologičeskogo obsluživanija [Automatized system for information-terminological service] - ASITO
Classification: Hierarchic classification conforming to the standards classification
Subject(s): standardization of terminology of various fields of science and technology. Elaboration of information retrieval systems for large terminological collections
UDC 0/9
Language(s): Ru; En; Fr; De
Service charges

○ *392* *(Organization)*

Suomen Standardisoimisliitto r. y.
[Finnish Standards Institution]
SFS
P. O. Box 205
Helsinki 12, SF-00121
Finland
Phone: 358 0 645 601
Telex: 122303 stand sf
Telegr.: finnstandard
Head: Mr. Kari Kaartama, Director
Vocabulary standards:
Subject(s): all fields
UDC 0/9
Language(s): Fi

● *393* *(Organization)*

Tekniikan Sanastokeskus
[The Finnish Centre for Technical Terminology (not official); Centralen fór Teknisk Terminologi (Sv)]
TSK
Parent organization: Kotimaisten kielten tutkimuskeskus
Liisankatu 16 B 13
Helsinki 17, SF-00170
Finland
Phone: 669 529; 179 161
Staff: 2
Head: Ms. Heidi Suonuuti, M. Sc (Chem.Eng.)
Expert(s): Ms. Heidi Suonuuti
Staff categories:
1 information specialist(s)
Founded: 1947
Documentation:
bibliographic data
factographic data
Subject(s): technology
UDC 62
Language(s): Fi; Sv; En; De
Vocabularies:
Subject(s): technology (general)
UDC 62
Language(s): Fi; Sv; En; De
Vocabulary standards:
Subject(s): technology
UDC 62
Language(s): Fi
Training: graduate; postgraduate
Access to training: available to outside users
Curriculum: fundamentals of terminology; terminological principles; lexicographical methods
Publications: *TSK tiedottaa.* ISSN 0358-108X (Information Newsletter); Tekniikan sanastoja 1979. *Tekniska ordlistor. Tekniikan Sanastokeskus & Kotimaisten kielten tutkimuskeskus.* Helsinki 1979. 36 p. ISBN 951-46-4382-8 (Catalogue of about 500 dictionaries, which contain Finnish and/or Swedish);
Såhkôalan sanastoja. Eltekniska ordlistor. Tekniikan Sanastokeskus. Helsinki, 1980. 14 p. ISBN 951-95566-2-1 (Catalogue of electrotechnical dictionaries, which contain Finnish or Swedish)
Services: other: Term service
No service charges
User aids: TSK tiedottaa (Information Newsletter)
Operation:
manual
computer-assisted (beginning 1980-02-01)

(Project(s))

Responsible: Commission of terminology of maintenance of real-estate
Title: Kiinteistönhuoltosonastohanke [Project of terminology of maintenance of real-estate]
Principal investigator: Timo Lounela
Collaborators: 9
Scope and/or subject matter: Terminology of maintenance of real-estate
Source material: Laws; standards; dictionaries; journals; handbooks; research reports etc.
Publications: *Multilingual dictionary of maintenance of real-estate*

Responsible: Commission of fire service terminology
Title: Palontojunta-alan sanastohanke [Project of fire service terminology]
Project type: institutional; in progress
Principal investigator: Paavo Heikkonen
Collaborators: 6
Scope and/or subject matter: Fire service
Source material: Havelka, F. International technical fire service dictionary, Vienna 1971; laws; standards; handbooks; journals etc.
Publications: *Multilingual fire service dictionary*

Responsible: Terminology commission of workers' protection
Title: Pohjoismainen työsuojelusanastohanke [Project of workers' protection terminology in northern countries]
Project type: institutional; in progress
Principal investigator: Thelma Aro
Collaborators: 7
Scope and/or subject matter: Workers' protection
Source material: Laws; standards; professional journals; research reports; handbooks; dictionaries etc.
Publications: Multilingual dictionary of workers' protection

Responsible: Commission for preparing a guide for preparing standardized vocabularies
Title: Standardisoitujen sanastojen laatimisohjehanke [A guide for preparing standardized vocabularies]
Project type: institutional; in progress
Principal investigator: Elisa Stenvall
Collaborators: 4
Scope and/or subject matter: Guide for standardized vocabularies
Source material: Standards of ISO/TC 37; DIN etc.
Publications: A guide for preparing standardized vocabularies

Responsible: Commission for preparing a vocabulary of terminology

Title: Terminologian laatimishanke [Vocabulary of terminology]
Project type: institutional; in progress
Principal investigator: Elisa Stenvall
Collaborators: 4
Scope and/or subject matter: Vocabulary of terminology
Source material: ISO R 1087 etc.
Publications: Finnish and Swedish vocabulary of terminology

● *394* *(Data Bank)*
Tekniikan Sanastokeskus
[The Finnish Centre for Technical Terminology (not official); Centralen fôr Teknisk Terminologi (Sv)]
TSK
Parent organization: Kotimaisten kielten tutkimuskeskus
Liisankatu 16 B 13 — Phone: 669 529; 179 161
Helsinki 17, SF-00170
Finland
Founded: Planned

● *395* *(Organization)*
Nordisk språksekretariat
[Nordic Language Secretariat]
Drammensveien 158
P. O. Box 8107 — Phone: (02) 55 98 09
Oslo 1
Norway
Staff: 3
Head: Cand. philol. Ståle Løland
Staff categories:
1 terminologist(s) (theory); 1 linguist(s)
Founded: 1978
Documentation:
bibliographic data
factographic data
Language(s): No; Sv; Da; and other Nordic languages
Vocabularies:
Subject(s): computers; marketing; petroleum
UDC 681. 3; UDC 658. 8; UDC 665. 5
Language(s): En; No; Sv; Da; and other Nordic languages
Vocabulary standards:
Subject(s): computers; marketing; petroleum
UDC 681. 3; UDC 658. 8; UDC 665. 5
Language(s): En; No; Sv; Da; and other Nordic languages

○ *396* *(Organization)*
Norges Standardiserigsfôrbund
[Norwegian Standards Association]
NSF
Haakon VII's gate 2 — Phone: 47 2 41 68 20
Oslo 1 — Telex: 19050 nsf n
Norway — Telegr.: standardisering
Head: Mr. Gudbrand Jenssen, Director
Vocabulary standards:
Subject(s): all fields
UDC 0/9
Language(s): No

● *397* *(Organization)*
Rådet for teknisk terminologi
[Norwegian Centre of Technical Terminology]
RTT
Riddervolds gate 3 — Phone: (02) 56 77 50
Oslo
Norway
Staff: 4
Head: Mr. Bernhard Christiansen, secretary general
Expert(s): Mr. Bernhard Christiansen
Staff categories:
1 translator(s); 3 other professional(s)
Founded: 1956
Documentation: factographic data
Subject(s): engineering (all fields)
UDC 62
Language(s): No; Da; En; Fr; Sv
Vocabularies:
Subject(s): engineering
UDC 62
Language(s): No; Da; En; Fr; Sv

Publications: The glossaries of RTT are multilingual with full cross reference possibilities between any combination of the languages mentioned below. The glossaries include succinct but clear-cut definitions in Norwegian. The glossaries are published and sold by Universitetsforlaget (University Press), P. O. B. 57, 5014 Bergen, Norway.
RTT 1–15 are obsolete.
RTT 16 *Glossary of work studies.* 203 Norwegian terms + En, Sv, De. 1968;
RTT 17 *Glossary of wood products* - Pulp and Cellulose. 510 Norwegian terms + En, Sv, De. 1968;
RTT 18 *Glossary of rock blasting technics.* 80 Norwegian terms + En, De. 1968;
RTT 19 *Glossary of dams* (no def). 1800 Norwegian terms + En, Fr, De, Es, It, Pt. 1968 (from ICOLD: *Technical Dictionary on Dams)*;
RTT 20 *Glossary of water supply and sewer systems.* Part 1 - Water supply. 1080 Norwegian terms + En, Fr, De. 1969;
RTT 21 *Glossary of plastics.* 1180 Norwegian terms + En, Fr, De. 1968;
RTT 22 *Glossary of railway technics* (no def.) 1995 Norwegian terms + Fr. 1969 (from UIC's *Lexique général des termes ferroviaires)*;
RTT 23 *Glossary of machine elements.* Part 1 - Bearings. 170 Norwegian terms + En, Sv, Dr, Fr. 1970;
RTT 24 *Glossary of machine elements.* Part 2 - Gears. 230 Norwegian terms + En, Sv, De, Fr. 1971;
RTT 25 *Glossary of rubber.* 1140 Norwegian terms + En, Sv, De. 1970;
RTT 26 *Glossary of building and real estate administration.* 211 Norwegian terms. 1969;
RTT 27 *Glossary of electric energy.* Part 1– Electric rotating machines. 900 Norwegian terms + En, Fr, De. 1972;
RTT 28 *Glossary of electric energy.* Part 2 - Transformers. 482 Norwegian terms + En, Fr, De. 1971;
RTT 29 *Glossary of electric energy.* Part 3 - Switchgear & Control gear. 332 Norwegian terms + En, Fr, De. 1973;
RTT 30 *Glossary of radiotechnology.* 2940 Norwegian terms + En, De. 1971/72;
RTT 31 *Glossary of road and traffic.* 1201 Norwegian terms + En, Sv, De, Da, Fr. 1973;
RTT 32 *Glossary of electric energy.* Part 4 - Installation equipment. 196 Norwegian terms + En, Fr, De. 1973;
RTT 33 *Glossary of ventilation.* 1132 Norwegian terms + En, Sv, De, Fr. 1974;
RTT 34 *Glossary of electric energy.* Part 5 - Generation, transmission and distribution of electric energy. 712 Norwegian terms + En, Fr, De. 1975;
RTT 35 *Glossary for the petroleum industry. Drilling and production. Geophysics, Geology.* 1550 Norwegian terms + En, Fr, 1976;
RTT 36 *Glossary of electric energy.* Part 6 - Static convertors. 1041 Norwegian terms + En, Fr, De. 1976;
RTT 37 *Glossary of electric energy.* Part 7 - Electric relays. 481 Norwegian terms + En, Fr, De. 1976;
RTT 38 *Glossary of water supply and sewer systems.* New edition. 2391 Norwegian terms + En, Fr, De. 1977;
RTT 39 *Glossary of automatic control.* 15 35 Norwegian terms + En, De. 1977/78;
RTT 40 *Glossary of telecommunications.* 2400 Norwegian terms + En, Fr, Sv, De. 1980;
RTT 41 *Glossary of alarm systems.* 290 Norwegian terms + En, Sv, De. 1980;
RTT 42 *Glossary of heating ventilation and sanitary technics.* 2300 Norwegian terms + En, Fr, Sv, De. 1980;
RTT 43 *Glossary of petroleum industry.* 1150 Norwegian terms + En. 1980.
Note: All glossaries have definitions in Norwegian.

Services: reference
No service charges

Operation:
computer-assisted

(Project(s))
Title: Avfallshåndtering [Waste management]
Project type: institutional; in progress
Principal investigator: Mr. Bernhard Christiansen
Collaborators: 3
Method of work: Computer assisted

Title: Bygningsglass [Building glass]
Project type: institutional; in progress
Principal investigator: Mr. Bernhard Christiansen
Collaborators: 3
Method of work: Computer assisted

Title: Fluidteknikk [Fluid technics]
Project type: institutional; in progress
Principal investigator: Mr. Bernhard Christiansen
Collaborators: 3
Method of work: Computer assisted

Title: Kartteknikk [Map technics]
Project type: institutional; in progress
Principal investigator: Mr. Bernhard Christiansen
Collaborators: 16
Method of work: Computer assisted

Title: Korrosjon [Corrosion]
Project type: institutional; in progress
Principal investigator: Mr. Bernhard Christiansen
Collaborators: 4
Method of work: Computer assisted

Title: Plast/gummi [Plastic/rubber]
Project type: institutional; in progress
Principal investigator: Mr. Bernhard Christiansen
Collaborators: 4
Method of work: Computer assisted

Title: Sanering og rehabilitering av VA-anlegg [Rebuilding and rehabilitation of water supply and sewers systems]
Project type: institutional; in progress
Principal investigator: Mr. Bernhard Christiansen
Collaborators: 6
Method of work: Computer assisted

Title: Vannturbiner [Water turbines]
Project type: institutional; in progress
Principal investigator: Mr. Bernhard Christiansen
Collaborators: 4
Method of work: Computer assisted

Title: Varmekraft: Kjelanlegg, turbinanlegg, raktoranlegg [Thermal power: Steam boiler for power plants, steam turbine installation, reactor installation]
Project type: institutional; in progress
Principal investigator: Mr. Bernhard Christiansen
Collaborators: 12
Scope and/or subject matter: Terminology relating to thermal power, such as steam boiler, steam turbines reactor installations, etc.
Method of work: Computer assisted

● ***398*** *(Data Bank)*
Norsk termbank
[The Norwegian Term Bank; Banque terminologique norvégienne]
NOTE
Parent organization: Rådet for teknisk terminologi
Strømgaten 53 — Phone: 47 5 320040, ext. 2952
Bergen, N-5000
Norway
Staff: 2
Head: Prof. Holbjørn Heggstad
Expert(s): Mr. Bjarne Norevik; Mr. Håvard Hjulstad
Staff categories:
2 linguist(s) / information specialist(s)
Founded: 1979
Purpose of data bank/users: Production of terminological dictionaries - both on paper and in machine-readable form; Mainly subject specialists
Operation:
computerized
Data base description:
Name: Norsk termbank
Subject(s): technical fields
UDC 5/7
Language(s): No; En; De; Fr; Sv; Da; Fi; Is
Terminological data:
definitions; classified layout (planned, but not (yet) fully included); context (sometimes); equivalents (varying, mostly En, De, Fr, some Sv, Da, Fi, Is)
Products: RTT 40 "Ordbok for teleteknikk", 1980.
RTT 41 "Ordbok for varsling og alarmteknikk", 1980;
RTT 42 "Ordbok for VVS", 1980;
RTT 43 "Ordbok for petroleumsvirksomhet", 1980;
RTT 44 "Ordbok for bygningsglass", 1982;
RTT 45 "Ordbok for vannturbiner", 1982;
RTT "Sanering og rehabilitering" av UA-nett, 1981;
RTT "Risikoanalyse", 1981; "Norsk Dataordbok", 1980; "Norsk Markedsføringsordbok", 1980; magnetic tapes, disks (on demand)
Service charges (by the hour and machine costs)
Access:
via telephone
on-line ; off-line (will be offered)
Data available:
with charge
Data processing:
Hardware/Capacity: UNIVAC 1100/80; magnetic tape; Various types of printers; Approx. 4 terminals through a concentrator
Software:
operating system: EXEC; operation mode: demand; programming language(s): mainly: Fortran, Pascal

● ***399*** *(Data Bank)*
The University of Oslo, Department of English; The Norwegian Ministry of Foreign Affairs
P. O. Box 1003 — Phone: 456892
Blindern, Oslo 3
Norway
Staff: 6 collaborators on the term bank project
Head: Mr. P. Chaffey, University of Oslo; Mr. O. Sandal, Ministry of Foreign Affairs
Founded: 1982
Purpose of data bank/users: To assist translators in the state administration; Translators, students
Data base description:
Name: Det norske termbankprosjektet (The Norwegian term bank project)
Subject(s): all fields
UDC 0/9
Language(s): No; En
Terminological data: in planning

○ ***400*** *(Organization)*
Standardiseringskommissionen i Sverige
[Swedish Commission of Standardization]
SIS
Tegnérgatan 11
Box 3 295 — Phone: 46 8 23 04 00
Stockholm, S-103 66 — Telex: 17453 sis s
Sweden — Telegr.: standardis
Head: Mr. Jan Ollner, Director
Vocabulary standards:
Subject(s): all fields
UDC 0/9
Language(s): Sv

● ***401*** *(Organization)*
Tekniska nomenklaturcentralen
[Swedish Centre for Technical Terminology; Centre suédois de terminologie technique; Schwedische Zentralstelle für Technische Terminologie (De)]
TNC
P. O. Box 5243 Phone: 46–8 24 92 90
Stockholm, S-102 45
Sweden
Visiting address: Brahegatan 38
Staff: 10
Head: Mr. Lars Algotsson
Expert(s): Mr. Göran Walter
Staff categories:
6 terminologist(s) (special field); 1 linguist(s); 1 information specialist(s); 2 other professional(s)
Founded: 1941
Documentation:
bibliographic data
factographic data
Language(s): Sv
Vocabularies:
Subject(s): technical fields (mainly)
UDC 5/7
Language(s): Sv; En; Fr; De; Es; Da; No; Fi; Ru (Definitions only is Sv, occasionally also in En)
Theory: Sv
Training: undergraduate; graduate; postgraduate
Access to training: available to outside users
Curriculum: fundamentals of terminology; terminological principles; lexicographical methods
Publications: *Glossary of Enamelling*, Sv - En - De, 1957, 850 terms, 104 p.; *Glossary of Plastics*, Sv - En - Fr - De, 1958, 1. 460 terms, 173 p.; *Glossary of Rubber*, Sv - En - Fr - De, 1968, 1350 terms, 147 p.; *Glossary of House-Painting*, Sv - En, 1968, 1. 210 terms, 140 p.; *Glossary of Water* 2, Sv - En - Fr - De - Fi, 1970, 340 terms, 80 p.; *Glossary of Concrete*, Sv - En - Fr - De - Da - No - Fi, 1971, 1. 310 terms, 366 p.;
Glossary of Brewing, Sv - En - Fr - De, 1971, 1. 200 terms, 204 p.; *Glossary of Petroleum*, Sv - En - Fr, 1971, 150 terms, 78 p.; *Glossary of Industrial Engineering*, Sv - En - De, 1971, 110 terms, 100 p.; *Glossary of Environment*, Sv - En - Fr - De, 1972, 780 terms, 90 p.; *Glossary of Astronautics*, Sv - En - Fr - De - Ru, 1973, 510 terms, 171 p.; *Glossary of Nuclear Energy*, Sv - En - Fr - De, 1975, 1. 400 terms, 396 p.; *Swedish Nomenclature of Inorganic Chemistry*, Sv - En, 1975, 380 terms, 118 p.; *Glossary of Heat Treatment*, En - Fr - De - Sv - Ru - Ja, 1974, 200 terms, 96 p.; *Glossary of Geotechnics*, Sv - En - Fr - De - Fi - Ru, 1975, 503 terms, 176 p.;
Glossary of Timber Construction, Sv - En - Fr - De, 1975, 725 terms, 224 p.; *Glossary of Municipal Engineering*, Sv - En - Fr- De - Da - No - Fi, 1976, 472 terms, 214 p.; *Glossary of Waste Management*, Sv - Da - No - Fi - En - Fr - De, 1977, 538 terms, 210 p.;
Glossary of Water Supply and Sewer, Sv - En - Fr- De - Da - Fi, 1977, 792 terms, 173 p.; *Glossary of Corrosion*, Sv - En - Fr - Es - De - Da - No - Fi - Ru, 1977, 540 terms, 235 p.;
Glossary of Strength of Materials, Sv - En - Fr - De - Da, 1977, 850 terms 230 p.; *Glossary of Air Treatment*, Sv - En - Fr - De - Fi, 1978, 724 terms, 204 p.;
Glossary of Glass in Building, Sv - En - Fr - De - Da - No - Fi, 1978, 274 terms, 106 p.;
Glossary of Forestry, Sv - En, 1978, 4. 570 terms, 676 p.; *Construction Vocabulary*, Sv-En, En-Sv, 1979, 9. 500 terms, 360 p.; *Glossary of Rock Engineering*, Sv - En - Fr - Es - De - Da - No - Fi, 1979, 800 terms, 306 p.; *Glossary of Paper*, Sv - En - Fr - De - No - Fi, 1979, 1233 terms, 416 p.; *Glossary of Planning and Building Terms*, Sv - En - Fr - De, 1980, 970 terms, 282 p.; *Glossary of Textiles*, Sv - En - Fr - De - Da - No - Fi, 1981, 780 terms, 272 p.
Note: All glossaries have definitions in Sv, except the *Glossary of Environment*.
Services: reference; loan; other: Telephone service for terminological queries
No service charges
User aids: *TNC Aktuellt*
Operation:
computer-assisted
Data base description:
Name: TERMDOK
Size: 70 000 records
Language(s): Sv; En; Fr; De; Es; Da; No; Fi; Ru
Classification: Internal
Terminological data:
definitions; explanations; classified layout; equivalents; other data elements
Data processing:
Hardware: Internal: Datapoint; external: Digital; Internal: DP 2200; external: DEC-system-10; Internal: 12 Kb; external: 1, 3 Mb; Internal: 2, 5 Mb; external: 90 Mb; Internal: cassette; external: 1/2 inch ASCII or EBCDIC 9 channel; internal: belt printer; serial word processing printer; External: various; External: all teletype-compatible terminals, 300 or 1200 baud full duplex, direct connection or by Nordic data network
Software:
operating system: Internal: DOS. A 2. 4; external: TOPS-10; operation mode: Internal: real time; external: real time, batch; programming language(s): Internal: Assembler, Databus 11; external: Assembler; data base management system: Internal: internal Assembler-written system; external: 3RIP
Access:
via telephone
on-line
Data available:
on exchange basis; with charge

● ***402*** *(Data Bank)*
Tekniska nomenklaturcentralen
[Swedish Centre for Technical Terminology; Centre suédois de terminologie technique; Schwedische Zentralstelle für Technische Terminologie (De)]
TNC
P. O. Box 5243 Phone: 46–8 24 92 90
Stockholm, S-102 45
Sweden
Visiting address: Brahegatan 38
Staff: 10
Head: Mr. Lars Algotsson
Expert(s): Mr. Göran Walter
Staff categories:
6 terminologist(s) (special field); 1 linguist(s); 1 information specialist(s); 2 other professional(s)
Purpose of data bank/users: A means for more efficient elaboration and dissemination of terminology; Technical writers, subject specialists, standards experts, journalists; translators
Data base description:
Name: TERMDOK
Size: 70 000 records
Language(s): Sv; En; Fr; De; Es; Da; No; Fi; Ru
Classification: Internal
Subject(s): technical fields
UDC 5/7
Language(s): Sv; En; Fr; De; Es; Da; No; Fi; Ru
Terminological data:
definitions; explanations; classified layout; equivalents; other data elements
Services: Telephone service
Access:
via telephone
on-line
Data available:
on exchange basis; with charge
Data processing:
Hardware/Capacity: internal: Datapoint; external: Digital internal: DP 2200; external: DEC-system-10; internal: 12 Kb; external: 1, 3 Mb; internal: 2, 5 Mb; external: 90 Mb; internal: cassette; external: 1/2 inch ASCII or EBDIC 9 channel; internal: belt printer; serial word processing printer; external: various; external: all teletype-compatible terminals, 300 or 1200 baud full duplex, direct connection or by Nordic data network

Software: operating system: internal: DOS. A 2. 4; external: TOPS-10; operation mode: internal: real time ; external: real time, batch; programming language(s): internal: Assembler, Databus 11; external: Assembler; data base management system: internal: internal Assembler-written system; external: 3RIP

○ ***403***(*Organization*)
Dansk Standardiseringsraad
[Danish Standards Council]
DS
Aurehøjvej 12
Postbox 77 Phone: 45 1 62 93 15
Hellerup, DK-2900 Telex: 15615 dansta dk
Denmark Telegr.: danskstandard
Head: Mr. Leif Noergaard, Director
Vocabulary standards:
Subject(s): all fields
UDC 0/9
Language(s): Da

● ***404***(*Organization*)
Terminologiafdelingen
[Department of Terminology (not official)]
Parent organization: Handelshøjskolen i København
Fabrikvej 7 Phone: (01) 19 19 19
København, DK-2000
Denmark
Visiting address: As above
Staff: 11
Head: Board of directors: Mr. Gert Engel; Ms. Inge Gorm Hansen; Mr. Heribert Picht; Mr. Jacques Qvistgaard. Co-ordinator and administrator: Ms. Bente Kristensen
Founded: 1974
Documentation:
bibliographic data
factographic data
Vocabularies:
Subject(s): agriculture; forestry; road construction; ergonomics; environment
UDC 63; UDC 634; UDC 625. 74; UDC 574; UDC 331. 01
Language(s): Da
Vocabulary standards: Part of ISO/TC 37, especially ISO 704
Subject(s): theory of terminology
UDC 001. 4. 001
Language(s): En
Theory: Da; De; En; Es
Training: undergraduate; graduate; postgraduate
Access to training: Limited
Curriculum: fundamentals of terminology; terminological principles; lexicographical methods
Publications: *NYT fra Terminologiafdelingen*, Handelshøskolen i København, nos. 1–6 (6–June 1980); CEBAL no. 5, *Special Issue on Terminology*. Copenhagen: Nyt Nordisk Forlag Arnold Busck, 1979; SPRINT, *Sproginstituternes Tidsskrift*, Handelshøjskolen i København, 1977–1981, total 12 numbers
User aids: NYT fra Terminologiafdelingen, Handelshøjskolen i København, nos. 1–6 (6–June 1980)

(*Project(s)*)
Title: Arbejderbeskyttelsesglosar [Glossary on Protection of Workers]
Project type: commissioned project (institutional Scandinavian Working Group; coordinator: Radet fôr Teknisk Terminologi (RTT), Oslo, Norway; in progress
Principal investigator: Mr. Gert Engel
Collaborators: 2
Scope and/or subject matter: Elaboration of terms and definitions in Da, En, No, Sv, Fi
Source material: Dictionaries, standards, text-books

● ***405***(*Data Bank*)
DANTERM - The Danish Terminological Data Bank
Parent organization: Handelshøjskolen i København
Fabrikvej 7
København, DK-2000
Denmark
Head: Mr. Gert Engel
Founded: In planning

● ***406***(*Organization*)
Bureau Terminologie en Documentatie
[Terminology and Documentation Section; Bureau de terminologie et de documentation; Referat Terminologie und Dokumentation (De) ; Oficina de terminología y documentación (Es)]
AVT/TD; AVT/TD; AVT/TD
Parent organization: Ministry of Foreign Affairs, Translations Branch
Casuariestrate 16
P. O. Box 20061 Phone: 70 - 209270
's-Gravenhage, 2500 EB Telex: 31326
Netherlands Telegr.: Celer
Visiting address: Dr. v. d. Stamstraat 2, Leidschendam
Staff: 11
Head: Mr. B. F. Tangelder, Drs.
Expert(s): Mw. Fabbricotti (Sv, It); S. Kleemann (Sv); Mw. Bourgeois Sv, De); F. v. Ruth (Sv, It)
Staff categories:
4 terminologist(s) (theory and special field); 4 documentalist(s)/librarian(s); 2 other professional(s)
Founded: 1070
Documentation:
bibliographic data
factographic data
Subject(s): all fields
UDC 0/9
Language(s): Nl; En; Fr; De
Vocabularies: Since 1980 co-operation with the Central Vertaalbureau Philips and the Vertaalgroep PTT concerning elaboration of joint Nl-En/Fr/De glossaries in all fields.
Theory: Marginal; Nl; De; En; Fr; Es
Training: Practical training for students and terminologists from other services.
Publications: Until end of 1981 products destined for internal usage only
Services: reference ; loan (restricted); translation (limited); referral (limited)
No service charges (usually)
Operation:
manual
computer-assisted (using equipment of Siemens Sprachendienst)
Data base description: Same as Siemens Sprachendienst
Terminological data:
definitions; explanations; classified layout; context; equivalents; other data elements
Access:
via telephone / via telex
off-line
Data available:
on exchange basis; free of charge (to be arranged beforehand)

(*Project(s)*)
Title: Financieel-economische aspecten ontwikkelingssamenwerking [Development aid: financial and economic aspects]
Project type: Internal usage; in progress
Collaborators: 5
Method of work: Collection and evaluation of sources in source language (Nl) and target languages (De, En, Fr) ; team-work (documentalists, terminologists, specialists in the field); common checking
Source material: Nl, En, Fr and De authentic sources. Nl experts in the field. Documents and expert knowledge offered by Bureau de Terminologie Bruxelles

Title: Onderwijsterminologie Nederland [Education terminology in the Netherlands]

Project type: Internal usage; in progress
Collaborators: 5
Method of work: Collection and evaluation of sources in source language (Nl) and target languages (De, En, Fr); team-work (documentalists, terminologists, specialists in the field); common checking
Source material: Nl sources, Nl experts in the field
Title: Vreemdtalige benamingen Nederlandse Organisaties [Translation of names of Dutch organizations]
Project type: Internal usage; in planning
Collaborators: 5
Method of work: Collection and evaluation of sources in source language (Nl) and target languages (De, En, Fr); team-work (documentalists, terminologists, specialists in the field); common checking
Source material: Nl sources

● ***407*** *(Data Bank)*
Bureau Terminologie en Documentatie
[Terminology and Documentation Section; Bureau de terminologie et de documentation; Referat Terminologie und Dokumentation (De); Oficina de Terminología y Documentatión (Es)]
AVT/TD; AVT/TD; AVT/TD
Parent organization: Ministry of Foreign Affairs, Translations Branch
Casuariestrate 16
P. O. Box 20061 — Phone: 70 - 209270
's-Gravenhage, 2500 EB — Telex: 31326
The Netherlands — Telegr.: Celer
Staff: 11
Head: Mr. B. F. Tangelder, Drs.
Expert(s): A. Bothe (De, En, Da, Sv); Mw. Fabbricotti (Sv, It); S. Kleeman (Sv); Mw. Bourgeois (Sv, De); F. v. Ruth (Sv, It)
Staff categories:
4 terminologist(s) (theory and special field); 4 documentalist(s)/librarian(s); 2 other professional(s)
Operation: computer-assisted
Subject(s): all fields
UDC 0/9
Language(s): Nl; En; Fr; De; It; Es

● ***408*** *(Organization)*
Centraal Vertaalbureau N. V. Philips Gloeilampenfabrieken
[Central Translation Department N. V. Philips]
Philips Eindhoven
Vestdijk 2d
Eindhoven
The Netherlands
Staff: 30
Head: Mr. K. W. Schneider, Head of Central Translation Dept.
Expert(s): Ms. Hoyer
Staff categories:
1 terminologist(s) (special field); 1 documentalist(s)/librarian(s); 18 translator(s); 10 other professional(s)
Documentation: factographic data
Subject(s): electronics; industrial engineering; business administration
UDC 621. 3; UDC 65
Language(s): Nl; En; Fr; Es; De
Services: translation
Service charges
Operation:
computer-assisted
Terminological data:
definitions; equivalents

○ ***409*** *(Organization)*
Nederlands Normalisatie-instituut
[Netherlands Standards Institute]
NNI
Kalfjeslaan 2
P. O. Box 5059 — Phone: 31 15 61 10 61
Delft, NL-2600 GB — Telex: 38 144 nni nl
Netherlands — Telegr.: normalisatie delft
Head: Dr. T. Dijs, Director
Vocabulary standards:
Subject(s): all fields
UDC 0/9
Language(s): Nl

○ ***410*** *(Organization)*
Shell Internationale Petroleum Maatschappij B. V., Translation Division
SIPM
Carel van Bylandtlaan 30 — Phone: 070–773373
's-Gravenhage
The Netherlands
Staff: 18 (16 full time, 2 part time)
Head: Mr. R. Rishworth
Expert(s): Mr. J. W. Schut
Staff categories:
13 translator(s); 2 other professional(s)
Documentation:
bibliographic data
factographic data
Subject(s): economics; chemistry; oil; gas
UDC 3/7
Language(s): En; Nl; De; Fr; Es; It
Vocabularies:
Subject(s): all fields; with special emphasis on economics, chemistry oil; gas
UDC 3/7
Language(s): En; Nl; De; Fr; Es; It
Services: translation (for in-house use only)

○ ***411*** *(Data Bank)*
Shell Internationale Petroleum Maatschappij B. V., Translation Division (SIPM/ODAH 1)
Carel van Bylandtlaan 30 — Phone: 070–773373
's-Gravenhage
The Netherlands
Head: Mr. R. Rishworth
Expert(s): Mr. J. W. Schut
Purpose of data bank/users: Support of internal translation activities; Members of Shell Translation Division in Europe
Subject(s): economics; chemistry; oil; gas; transportation; engineering; human environment
UDC 3/7
Language(s): En; Nl; De; Fr; Es; It
Terminological data:
definitions; explanations; context; standardized terms; equivalents
Products: Computerized dictionary with bi-annual printout of updated master, sorted by languages and with provision for future sorting by in-house defined subject fields

● ***412*** *(Organization)*
Stichting International Translations Centre
ITC
Doelenstraat 101 — Phone: 015–142242, 015–142243
Delft, 2611 NS — Telex: 38104
The Netherlands
Staff: 16
Head: D. van Bergeijk
Expert(s): D. van Bergeijk
Staff categories:
1 documentalist(s)/librarian(s); 1 information specialist(s); 1 translator(s); 13 other professional(s)
Founded: 1961
Documentation: bibliographic data
Subject(s): translation
UDC 82. 03
Language(s): all languages
Publications: *World Transindex* - published monthly; *Journals in Translation* - 1978; *Translation News* - irregular publication; *World Transindex* is available on-line via ESA/IRS timespan 1978
Services: referral
User aids: *Translation News*

● *413**(Organization)*

Centre de terminologie de Bruxelles

[Brussels' Center for terminology; Centrum voor Terminologie van Brussel (Nl)]

CTB; BTC (not official); CTB (Nl)

Parent organization: Institut Libre Marie Haps
rue d'Arlon 11 Phone: 513. 23. 86
Bruxelles, B-1040
Belgium
Visiting address: 1, place du Luxembourg, B-1040 Brussels
Staff: 25
Head: Mr. Georges Lurquin, Scientific adviser of the Institute Marie Haps
Expert(s): Ms. Caroline de Schaetzen
Staff categories:
3 terminologist(s) (special field); 4 terminologist(s) (theory); 2 documentalist(s)/librarian(s); 4 linguist (s); 10 translator(s); 11 terminology teacher(s); 1 other professional(s)
Founded: 1978
Documentation:
bibliographic data
factographic data
Subject(s): all fields
UDC 0/9
Language(s): Fr; En; De; Nl; Es; It; Ru; Ja; Gr
Vocabularies:
Subject(s): all fields
UDC 0/9
Language(s): Fr; Nl; Es; It; De; En
Vocabulary standards:
Subject(s): all fields
UDC 0/9
Language(s): Fr; Nl
Theory: Fr; Es
Training: graduate
Curriculum: fundamentals of terminology; terminological principles; lexicographical methods graduates: courses in the programme of the degree in translation (Marie Haps): 1. Lexicological exercises, 2. Initiation into scientific and technical neology, 3. Stages or dissertation in terminology;
Publications: Henri Van Hoof. *International bibliography of specialized dictionaries.* KG Saur Verlag München, 1979, 470 p.;
G. de Crayencour. *Dictionnaire héraldique.* Tous les termes du blason. Bruxelles, 1974, complément: 1976, Service de Centralisation des études généalogiques et démographiques de Belgique (SCGD);
G. Lurquin. *Le français; langue des affaires et terminologie*; Enseignement et gestion, pp. 47–57, July 1980; G. Lurquin. *Chronique du langage*, Data News, Bruxelles; G. Lurquin. *La formation des traducteurs, demain*, Le langage et l'Homme, no. 43, May 1980, pp. 67–76;
G. Lurquin et M. Vincent. *Terminologie et terminographie*, Le langage et l'Homme, no. 36, January 1978, pp. 45–59;
B. de Bessé. *Terminologie et traduction*, Le langage et l'Homme, no. 34, May 1977, pp. 18–28;
G. Lurquin. *La terminologie de l'investissement*, Intermédiaire, no. 14, July 1979, Bruxelles;
G. Lurquin. *Le lexique de la télématique*, Intermédiaire, no. 1, January 1980;
Centre de Terminologie de Bruxelles. *Les Cahiers de terminologie. Bulletin de bibliographie terminologique.* Fields already covered: génétique: recombinaison chez les bactéries et les virus; basket-ball; droit commercial: contrat de vente; obstétrique: grossesse et accouchement; médecine: psychopharmacologie clinique; professions: informatique; droit civil: droit successoral; construction: isolation thermique; cytogénétique: pathologie; travaux de peinturage; imprimeries: mise en page typographique; professions: ingénierie; ondes radiophoniques; horaire variable; énergie nucléaire: surrégénérateurs; sociétés anonymes; environnement: pollution par le le bruit; engins de levage; haute fidélité; droit civil: responsabilité civile; automobile: sécurité
1 000 cards a year:
Cahiers de terminologie, size DIN A 6, with an entry, a definition, a context, notes (+ ref., indexes) for 2 languages at least under which Fr, and an indication of the field. Each *Cahier* has also a bilingual glossary. The cards are intended to complete the documentation of the translators. The vocabularies over particular fields can be bought separately.
The *Cahiers* can be obtained free of charge on exchange basis. The *Bulletin de bibliographie terminologique* is included free of charge in the *Cahiers*.
Services: reference (bibliographies of multilingual dictionaries in any field); loan (books of the library of the Institute Marie Haps); translation (pinpoint consultation by mail or telephone) for difficult words for translators and for search of references); other: proposing of neologisms (e.g.for scientists); selling of terminological cards
Service charges: 1. 000 bfr a year for 1. 000 terminological cards
No service charges for other services
User aids: Abstracts of new dictionaries, colloques, books, in *Le langage et l'Homme*
Data base description:
Name: Fichier de terminologie
Size: 12 000 units
Language(s): Fr; En; De; Nl; Es; It (mostly Fr and En)
Terminological data:
definitions (with references); context; equivalents (in at least one other language); other data elements (linguistic, encyclopedical)

◐ *414**(Organization)*

Centre national de documentation scientifique et technique, Bibliothèque Royale Albert I

[National Centre of Scientific and Technical Documentation (not official)]

CNDST-BR

4, bd. de l'Empereur
Bruxelles, B-1000
Belgium
Staff: 44
Head: Dr. A. Cockx
Staff categories:
4 translator(s); 3 linguist(s) / terminologist(s) (special field); 37 documentalist(s) / librarian(s) / information specialist(s)
Founded: 1964
Documentation: bibliographic data
Publications: Different publications
Services: reference; reproduction; translation
Service charges: For certain services
Data processing:
Hardware: Siemens; 4004/135; 876 Kb

● *415**(Organization)*

Hoger Instituut voor Vertalers en Tolken van het Rijksuniversitair Centrum Antwerpen

[State University College for Translators and Interpreters (not official); Institut supérieur de traducteurs et interprètes du Centre Universitaire de l'Etat à Anvers]

H. I. V. T.

Schildersstraat 41 Phone: (031) 38. 98. 32
Antwerpen, B-2000
Belgium
Staff: 40
Head: Prof. D. Godfrind
Expert(s): Prof. Dr. Jos. De Cort
Staff categories:
1 terminologist(s) (theory); 4 documentalist(s) / librarian(s); 7 linguist(s); 1 terminology teacher(s)
Founded: 1961
Documentation: factographic data
Subject(s): all fields
UDC 0/9
Language(s): Nl; Fr; En; De; It; Es; Ru; Pt
Vocabularies:
Subject(s): all fields
UDC 0/9
Language(s): Nl; Fr; En; De; It; Ru; Es; Pt

Theory: De
Training: graduate
Curriculum: fundamentals of terminology; terminological principles
Publications: *Linguistica Antverpiensia*, issued since 1967, 1 vol. yearly, approx. 250 p. Publications in different languages, especially articles on applied linguistics and translation

○ ***416***(*Organization*)
Institut belge de normalisation
IBN
Av. de la Brabançonne, 29 Phone: 32 2 734 92 05
Bruxelles, B-1040 Telex: 23 877 benor
Belgium Telegr.: benor
Head: M. P. Croon, Directeur général
Vocabulary standards:
Subject(s): all fields
UDC 0/9
Language(s): Fr; Nl

○ ***417***(*Organization*)
Association suisse de normalisation
SNV
Kirchenweg 4
Postfach Phone: 41 1 47 69 70
Zürich, CH-8032 Telex: 54 942 vsm
Switzerland Telegr.: normbureau
Head: Dr. Hans C. Zürrer, Directeur
Vocabulary standards:
Subject(s): all fields
UDC 0/9
Language(s): De; Fr; It

● ***418***(*Organization*)
Institut für nichtnumerische Informationsverarbeitung
[Institute of Non-Numerical Information Processing; Institut pour le traitement de l'information non numérique; Instituto per l'elaborazione dell'informazione non numerica (It)]
INIV
Thaler Straße 8
P. O. Box 500 Phone: 071/42 36 38
Rorschach, CH-9400
Switzerland
Staff: free-lance collaborators
Head: Herbert E. Bruderer
Expert(s): Herbert E. Bruderer
Founded: 1976
Documentation: factographic data
Subject(s): computer science; information science; information processing
UDC 681. 3; UDC 007
Language(s): De; En
Vocabularies:
Subject(s): information processing
UDC 681. 3; UDC 007
Language(s): De; En
Publications: Bruderer, H. E. *Nichtnumerische Informationsverarbeitung*. Bodenseeverlag, CH-9400 Rorschach, 1979; Bruderer, H. E. *Nichtnumerische Datenverarbeitung*. Mannheim/Wien/Zürich: Bibliographisches Institut, 1980.
Services: translation (De; En; Fr; It); other: data processing consulting
Operation:
manual

(*Project(s)*)
Title: Wörterbuch der nichtnumerischen Informationsverarbeitung. Datenverarbeitung, Information und Dokumentation, Telematik. En - De; De - En [Dictionary on Non-Numerical Information Processing]
Project type: institutional; in progress
Principal investigator: Herbert E. Bruderer
Collaborators: 2
Scope and/or subject matter: Information processing ; computer science
Source material: All sorts of publications, reports, etc., esp. grey literature
Publications: Bruderer, H. E. *Nichtnumerische Informationsverarbeitung*. Bodenseeverlag, CH-9400 Rorschach, 1979; Bruderer, H. E. *Nichtnumerische Datenverarbeitung*. Mannheim Wien Zürich: Bibliographisches Institut, 1980.

● ***419***(*Organization*)
Schweizerische Zentralstelle für Baurationalisierung
[Swiss Research Centre for Building Rationalization; Centre suisse d'études pour la rationalisation du bâtiment; Centro svizzero di studio per la razionalizzazione dell'edilizia (It)]
CRB
Parent organization: BSA Bund Schweizer Architekten; SIA Schweizerischer Ingenieur- und Architekten-Verein; SBV Schweizerischer Baumeisterverband
Seefeldstraße 214 Phone: 01 - 55 11 77
Zürich, CH-8008 Telegr.: CRB Zürich
Switzerland
Staff: 11
Head: Dipl. Arch. Heinz Joss
Expert(s): Dipl. Arch. Heinz Joss
Founded: 1959
Documentation: bibliographic data
Subject(s): building
UDC 69; UDC 72
Language(s): De; Fr; It; (En)
Vocabularies:
Subject(s): building
UDC 69; UDC 72
Language(s): De; Fr; It; (En)
Vocabulary standards:
Subject(s): building
UDC 69; UDC 72
Language(s): De; Fr; It; (En)
Publications: More than 100 technical/administrative publications in De, some 90 in Fr, 30 in It; A dictionary: Frommhold; Gareiss. *Bauwörterbuch. Begriffsbestimmungen aus dem Bauwesen* (approx.5500 concepts). 2nd ed., Werner Verlag, Düsseldorf, 1978; *Bulletin CRB*, a journal published 3–4 times a year

○ ***420***(*Organization*)
Ellinikos organismos typopiiseos
[Hellenic Organization for Standardization]
ELOT
Didotou 15 Phone: 30 1 360 95 17
Athens 144 Telex: 21 9621 elot gr
Greece Telegr.: elotyp-athens
Head: Mr. Evangelos P. Vardakas, Managing Director
Vocabulary standards:
Subject(s): all fields
UDC 0/9
Language(s): Gr

● ***421***(*Organization*)
Kéndron Ekpedeftikón Meletón ké Epimorfóseos
[Centre for Educational Research and Teacher-in-Service Training/Centre de recherches en éducation et du perfectionnement des enseignants]
KEME
Parent organization: Ministry of National Education and Religion of Greece
396 Messoyion St. Phone: 6567364; 6567368
Athens, Ayia Paraskevi
Greece
Staff: 125
Head: Prof. Pavlos Sakellaridis, President of KEME (Ex-Rector, Athens Polytechnic School)
Expert(s): Dr. Spyros E. Diamessis, Counsellor of KEME
Staff categories:
1 terminologist(s) (special field); 1 terminologist(s) (theory); 1 documentalist(s)/librarian(s); 1 linguist (s); 1 information specialist(s); 2 translator(s); 1 terminology teacher(s); 3 other professional(s)
Founded: 1975/76

Documentation: bibliographic data
Subject(s): engineering; physics; mathematics; education; philosophy
UDC 1/6
Language(s): Gr; En
Vocabularies:
Subject(s): engineering; physics; mathematics; education; philosophy
UDC 62; UDC 53; UDC 51; UDC 37; UDC 1
Language(s): Gr; En
Vocabulary standards:
Subject(s): electrical engineering; physics; mathematics; education; philosophy
UDC 621. 3; UDC 53; UDC 51; UDC 37; UDC 1
Language(s): Gr; En
Theory: Gr; En
Publications: Published individually by 1 or 2 members of KEME in the past or presently in print, e. g. electronics dictionary; computers dictionary; mathematics dictionary; physics dictionary, etc. (available upon request)
Services: reproduction; translation
Service charges
User aids: Planned
Operation:
manual
computer-assisted
Data base description: Not fully developed yet
Terminological data:
standardized terms; equivalents
Data processing:
Software: Limited use of FORTRAN at an initial stage in the past. Not extensive work in this area
Access: Not developed yet
Data available:
on exchange basis; free of charge

(Project(s))
Responsible: Greek EUDISED Thesaurus Group
Title: Helliniki apodosi tou thesavrou EUDISED [Greek version of the EUDISED Thesaurus]
Project type: institutional; commissioned project (EEC-Council of Europe); in progress
Principal investigator: Dr. Spyros E. Diamessis, counsellor - KEME
Collaborators: 6
Scope and/or subject matter: Publication of a Greek version of the wellknown EUDISED multilingual thesaurus•(there already exist versions in En, Fr, De, Nl, and - more recently - Es and Pt). In view of Greece's imminent entry into the EEC (1/1/1981) this need is evident for sufficient communication and exchange of information in education (as well as other fields).
Method of work: (1) Translation of all the EUDISED thesaurus terms into Greek from, principally, the English (and French or German) original; (2) elaboration through cross-referencing and cross-checking by means of alternative versions (e.g.German version). Addition of possible new terms, if desired.
Source material: (1) Original text (En, Fr, De) of the thesaurus; (2) S. Diamessis' dictionaries in education, philosophy, etc. as sources for the terminological material ; (3) other dictionaries on the market; (4) linguistic material in En and Gr; (5) other terminology and thesaurus work available
Publications: The target publication is the thesaurus in Greek; no other (intermediate) publication will be available; background work on terminology already exists (copies available on complimentary basis)

Responsible: Technical and Technological Terminology Group
Title: Lexicó tehnicón ké tehnoloyikón oron [Dictionary of technical and technological terminology]
Project type: institutional; commissioned project; KEME and possibly Ministry of Coordination, Agency for Scientific Research and Development; in progress
Principal investigator: Dr. Spyros E. Diamessis, counsellor of KEME
Collaborators: 4–6
Scope and/or subject matter: Publication and distribution (free) of a Greek-English and English-Greek Dictionary of technical and technological terminology (with mathematical terms) to Greek schools for students (secondary level) of vocational education: fields: electrical engineering; electronics; mechanical engineering, civil engineering
Method of work: (1) Collection of terms from all the relevant technical school books; (2) enrichment with all other necessary technical terms; (3) comparative terminological study (Greek, English), selection of most accurate terms; (4) study-research of practical use of terms on job (industry, workshop, laboratory) for each case; (5) classification of terms, inclusion of mathematics and computer terminology; (6) printing of a number of copies for free distribution to schools; (7) possible extension of work to other fields and languages
Source material: (1) All the technical books (textbooks) from secondary technical-vocational education schools in Greece; (2) S. Diamessis' series of related dictionaries and glossaries in the aforementioned and other related fields; (3) publications and teaching notes of the other members of the group; (4) all other publications - linguistic, terminological - on the market (Greek, English, etc.)
Publications: Series of dictionaries and related publications and/or work by: S. E. Diamessis (engineering, mathematics, physics, etc.); A. Sipitanou (general education); D. Panayiotacos (computers); S. E. Diamessis' India Conference Paper on new *Transliteration Grammar*. Copies available upon request (free of charge - complimentary copies); Those of above authors available in draft form on request, e. g. education terminology (plus EUDISED Thesaurus draft material in Greek); vocational education terminology; work on machine translation between Indo-European languages

● ***422*** *(Commission/Committee)*
Omádha Tehnikis ké Tehnologhikis Orologhias
[Technical and Technological Terminology Group ("Scientific Terminology" Group); Equipe pour terminologie scientifique (technique et technologique)]
"O. TE. T. O."
Parent organization: Ministry of National Education and Religion of Greece, KEME (Center for Educational Research and Teacher-in-Service Training)
396 Messoyion St. Phone: 6567364; 6567368
Athens, Ayia Paraskevi
Greece
Contact: Dr. S. E. Diamessis
Room 315
KEME
396 Messoyion St.
Ayia Paraskevi
Athens
Greece
Committee members: 6
Chairperson: Dr. Spyros E. Diamessis, counsellor, KEME
Vice-chairperson: Mr. Demetrios Panayiotacos, Scientific collaborator from Ministry of Labour
Established: 1980 (unofficial previously)
Vocabularies:
Subject(s): engineering; mathematics
UDC 5/6
Language(s): Gr; En
Vocabulary standards:
Subject(s): engineering; mathematics (principally)
UDC 5/6
Language(s): Gr; En
Theory: Yes; Greek and English terminology in engineering, mathematics, among others; Gr; En
Publications: Published by 1 or 2 members of the group in the past, e. g. electronics dictionary; circuits dictionary ; transistors and semiconductors dictionary; mathematics dictionary, etc. (available upon request); Complementary copies

● *423* *(Data Bank)*
Kéndron Ekpedeftikón Meletón ké Epimorfóseos
Founded: In planning
Data base description: Not fully established and/or organized yet. The only related work: Paper presented at the 3rd International Study Conference on Classification Research, Bombay, India, 6–11 January 1975 (see proceedings entitled 'A comparative linguistics approach to classification and information processing: A computer grammar for Indo-European machine translation - transliteration based on (Latin and) Greek'), plus some of the dictionaries (English-Greek, Greek-English).

○ *424* *(Organization)*
Jugoslovenski zavod za Standardizaciju
[Yugoslav Institute of Standardization]
JZS
Slobodana Penezića-Krcuna br. 35
Pošt. Pregr. 933
Beograd, YU-11000
Yugoslavia
Phone: 38 11 64 35 57
Telex: 12089 yu yus
Telegr.: standardizacija
Head: Mr. Vukašin Dragojević, Director
Vocabulary standards:
Subject(s): all fields
UDC 0/9
Language(s): Sh

○ *425* *(Organization)*
Makedonska Akademija na Naukite i Umetnostite
MANU
Lermontova 1
P. O. Box 428
Skopje, YU-91001
Yugoslavia
Staff: 117 (116 part time, 1 full time)
Head: Mr. Blaže Koneski, Acad. (President)
Expert(s): Ms. Lidija Ežova
Staff categories:
9 linguist(s) /documentalist(s)/librarian(s)
Founded: 1967
Documentation: bibliographic data
Subject(s): all fields of science; technology; humanities; civil engineering; architecture; electrical engineering; chemistry; flora; fauna; sociology, and others
UDC 0/9
Language(s): Mc; Sh; Sn; En; De; Fr; Ru
Publications: *Bilten na Odborot za izrabotuvanje na makedonska terminologija* [Bulletin of the Committee for working out the Macedonian Terminology]. Skopje: MANU, 1971- 1975; Editor: L. Ežova

(Project(s))
[Compilation of "Macedonian terminology of all fields of science and knowledge"]

● *426* *(Organization)*
Referalni centar Sveučilišta u Zagrebu
[Referral centre of the University of Zagreb]
RC; RC
Parent organization: RO Bibliotečno informacijski centar Hvratske-OOUR
Trg maršala Tita 3
P. O. Box 327
Zagreb, YU-41001
Yugoslavia
Phone: 041/448-072, 446-330, 445-600, 413-864, 448-071
Telex: 22206 yu bich
Staff: 58
Head: Dr. Ivan Mihel
Staff categories:
6 documentalist(s)/librarian(s); 2 linguist(s); 12 information specialist(s); 5 translator(s); 8 other professional(s)
Founded: 1967
Documentation:
bibliographic data
factographic data
Subject(s): information handling
UDC 002; UDC 659. 2; UDC 007
Publications: Monthly current awareness list of standards available at RC; *Informatologia Yugoslavica* - INYUAL, ISSN 0046-9483; *IRCIHE Bulletin-* IRCE8S ISSN 0351-0123 (annually); *SCIENTIA YUGOSLAVICA*-SYU; ISSN 0350-686X
Services: reference; reproduction; referral
Service charges: For reproduction services only
User aids: Others
Operation:
computer-assisted
Data processing:
Hardware: UNIVAC; 1110; U-100, DCT-1000 (leased)
Software:
operating system: EXEC-8; operation mode: batch ; demand; programming language(s): ASSEMBLER, FORTRAN

● *427* *(Commission/Committee)*
Terminološka komisija SAZU
[Terminological Commission SASA [Slovene Academy of Science and Arts]; Commission terminologique ASAS [Académie des sciences et des beaux arts de Slovénie)]]
TK SAZU; TC SASA (not official)**; CT ASAS** (not official)
Parent organization: Slovenska akademija znanosti in umetnosti
Novi trg 3
Ljubljana, YU-61000
Yugoslavia
Phone: (061) 23-961; (061) 23-964
Committee members: 6
Vice-chairperson: Prof. dr. Janez Batis
Established: 1949
Vocabularies:
Subject(s): law; technology; medicine; veterinary medicine; natural sciences; arts
UDC 3/7
Language(s): Sn
Vocabulary standards:
Subject(s): technology
UDC 6
Language(s): Sn
Publications: *The yearbook of the Slovene Academy of Sciences and Arts* (without charge)

○ *428* *(Organization)*
B"lgarska Akademija na Naukite, Centralna Biblioteka
[Central Library of the Bulgarian Academy of Sciences]
1, Sedmi Noemvri
st. Sofia
Bulgaria
Staff: 61 (59 full time, 2 part time)
Head: Ms. Elena Savova
Staff categories:
26 linguist(s)/terminologist(s) (special field); 2 translator(s); 32 documentalist(s)/librarian(s); 10 other professional(s) (consultants)
Founded: 1869
Documentation: bibliographic data
Subject(s): social sciences; natural sciences; science policy
UDC 3; UDC 5
Language(s): Bg; Ru; En
Vocabularies:
Subject(s): social sciences; natural sciences; science policy
UDC 3; UDC 5
Language(s): Bg; Ru; En
Publications: Participation in elaboration of *Bibliography of Interlingual, Scientific and Technical Dictionaries*, Paris: Unesco, 1961, 263 p.; SAVOVA, E. *Naučnata informacija i nejnata terminologija.* In: Parvi Kongres na B"lgarskoto istoričesko drujestvo, Sofia, 1970, t. 2, Sofia 1972, p. 531-539
Services: reference; other: Library and documentation services
No service charges

○ ***429*** *(Organization)*
D"rzaven Komitet za Standartizacija pri Ministerskija S"vet
[State Committee for Standardization at the Council of Ministers]
DKS
21, 6th September Str. Phone: 85 91
Sofia Telex: 22570 dkc bg
Bulgaria Telegr.: dkc
Head: Mr. Nikola Kaltchev, President
Vocabulary standards:
Subject(s): all fields
UDC 0/9
Language(s): Bg

○ ***430*** *(Organization)*
Institutul Român de Standardizare
[Romanian Institute of Standardization]
IRS
Căsuţa Poştală 63–87 Phone: 33 76 60
Bucarest 1 Telex: 11312 cnst r
Romania Telegr.: stas
Head: M. Dumitru Mihai
Vocabulary standards:
Subject(s): all fields
UDC 0/9
Language(s): Ro

○ ***431*** *(Organization)*
China Association for Standardization
CAS
P. O. Box 820 Phone: 89 59 98
Beijing
China Telegr.: 0621 beijing
Head: Mr. Yue Zhijian
Vocabulary standards:
Subject(s): all fields
UDC 0/9
Language(s): Ch; En

● ***432*** *(Organization)*

[Hong Kong Standards and Testing Centre]
HKSTC
21A Ma Tau Wei Road,
12/F., Hung Hom, Kowloon Phone: 3–659061
Telex: 84652 hkind hx
Hong Kong Telegr.: FEDINDUSTR HONGKONG
Staff: 65
Head: Mr. C. S. O. Chan
Staff categories:
2 documentalist(s)/librarian(s); 9 information specialist(s); 4 translator(s)
Founded: 1960
Documentation: bibliographic data
Subject(s): commerce; industry
UDC 33; UDC 6
Language(s): En; Ch
Vocabulary standards: Adoption of international and/or national terminological standards
Subject(s): commerce; industry
UDC 0/9
Publications: *Hong Kong Industrial News*
Services: reference; loan; reproduction; translation; referral
Service charges
Operation:
manual
Access:
via telephone/via telex
Data available:
on exchange basis; with charge

○ ***433*** *(Organization)*
Committee for Standardization of the Democratic People's Republic of Korea. Committee of the Science and Technology of the State
CSK
Sosong guyok Ryonmod dong
Pyongyang
Korea, Dem. P. Rep. of Telegr.: standard
Head: Mr. Li Gon Sik, President
Vocabulary standards:
Subject(s): all fields
UDC 0/9
Language(s): Ko; En

○ ***434*** *(Organization)*
Bureau of Standards. Industrial Advancement Administration
KBS
Yongdeungpo-Dong Phone: 633 95 59
Seoul
Korea, Rep. of
Head: Mr. Kye-Mook Jun, Director General
Vocabulary standards:
Subject(s): all fields
UDC 0/9
Language(s): Ko; En

● ***435*** *(Organization)*
Denki Gakkai
[Institute of Electrical Engineers of Japan]
Yurakucho 1–12–1,
Chiyoda-ku Phone: 201–0983
Tokyo, 100
Japan
Staff: 25
Expert(s): Mr. Hiroshi Miyakawa
Staff categories:
other professional(s)
Documentation:
bibliographic data
factographic data
Subject(s): electrical engineering
UDC 621. 3
Language(s): Ja
Vocabularies:
Subject(s): electrical engineering
UDC 621. 3
Language(s): Ja
Vocabulary standards:
Subject(s): electrical engineering
UDC 621. 3
Language(s): Ja

(Project(s))
Responsible: Special Standards Committee of Electrical Engineering Terminology
Title: Hyōjun Yōgo (Denki Kōgaku); Senmon Yōgo (Denki Kōgaku) [Standard Terminology (electrical engineering); special terminology (electrical engineering)]
Project type: institutional; in progress
Principal investigator: Mr. Hiroshi Miyakawa
Collaborators: About 30 committee members
Scope and/or subject matter: terminology of electrical engineering (in Ja)
Source material: IEC vocabulary, IEEE dictionary, text books, handbooks
Publications: *Standard Terminology*. About 10 booklets on Standard Terminology in special fields have been published already.

● ***436*** *(Commission/Committee)*
Denki Gakkai Denki Yōgo Hyōjun Tokubetsu Iinkai
[Special Standards Committee on Electrical Engineering

Terminology of the Institute of Electrical Engineers of Japan]
Parent organization: Institute of Electrical Engineers of Japan
Yurakucho 1–12–1,
Chiyoda-ku Phone: 201–0983
Tokyo, 100
Japan
Contact: Mr. Hiroshi Miyakawa
Faculty of Engineering
University of Tokyo
Bunkyo-ku, Hongo 7–3–1, Tokyo, Japan
Committee members: 30
Chairperson: Mr. Hiroshi Miyakawa
Vocabularies:
Subject(s): electrical engineering
UDC 621. 3
Language(s): Ja
Vocabulary standards:
Subject(s): electrical engineering
UDC 621. 3
Language(s): Ja

○ ***437*** *(Organization)*
Japanese Industrial Standards Committee. Standards Department. Agency of Industrial Science and Technology. Ministry of International Trade and Industry
JISC
1–3–1, Kasumigaseki,
Chiyodaku Phone: 81 3 434 59 15/8
Tokyo 100
Japan Telegr.: mitijisc tokyo
Head: Mr. Renzo Taguchi, President
Vocabulary standards:
Subject(s): all fields
UDC 0/9
Language(s): Ja; En

● ***438*** *(Organization)*
Kokusai Igaku Jōho Sentā
[International Medical Information Center]
IMIC
Parent organization: The Ministry of Health and Welfare; The Ministry of Education
26 Daikyo-cho Phone: 03–357–9181;
Shinjuku-ku, Tokyo, 160 03–357–8774
Japan Telex: 2323141 IMICJP J
Staff: 62 full time, 30 part time
Head: Dr. Daizo Ushiba, M. D., Chairman, Board of Directors
Staff categories:
8 terminologist(s) (special field); 18 documentalist(s)/librarian(s); 21 information specialist(s); 4 translator(s); 10 other professional(s)
Founded: 1972
Documentation:
bibliographic data
factographic data
Subject(s): biomedical sciences; health sciences
UDC 57; UDC 61
Language(s): En; Ja
Vocabularies:
Subject(s): biomedical sciences; health sciences
UDC 57; UDC 61
Language(s): En; Ja
Training: Training possibilities are offered in the field of biological and medical terminology in En and Ja
Curriculum: fundamentals of terminology
Services: reference; loan; reproduction; translation; referral; SDI; other: Abstracting, literature search and analysis
Service charges
User aids: IMIC (International Medical Information Center)
Operation:
manual
computer-assisted
Data base description:
Name: JOIS-JICST
Data processing:
Software: retrieval documentation system: JOIS-JICST on-line Information System
Access:
via telephone/via telex

(Project(s))
Title: International Cancer Research Data Bank
Project type: institutional; in progress
Principal investigator: Toshio Toyama, M. D., Managing Director
Collaborators: 10
Scope and/or subject matter: Development and Implementation of Mechanisms for Interaction between the ICRDB Program and Asian Countries
Method of work: Screening original journals and proceedings, and making abstracts of their articles
Source material: Journals, dictionaries, handbooks, directories, etc.
Publications: Data will be available with CANCERLINE (NCI, U.S.A.)

● ***439*** *(Organization)*
Monbushō Gakujutsu-kokusaikyoku Jōhō-toshokanka
[Science Information and University Library Division, Science and International Affairs Bureau, Ministry of Education, Science and Culture]
2–2, Kasumigaseki 3–chome,
Chiyoda-ku Phone: 03–581–4211
Tokyo, 100
Japan
Staff: 3
Head: Mr. Akira Tabohashi, Director, Science Information and University Library Division
Expert(s): Mr. Kunio Aoto, Officer for Science Affairs, Science and International Affairs Bureau
Staff categories:
1 terminologist(s) (special field); 2 other professional(s)
Founded: 1871
Documentation:
bibliographic data
factographic data
Subject(s): all fields
UDC 0/9
Language(s): Ja; En
Vocabularies:
Subject(s): all fields (23)
UDC 0/9
Language(s): Ja; En
Vocabulary standards:
Subject(s): all fields (34)
UDC 0/9
Language(s): Ja; En
Publications: Scientific terms in 23 branches of science and technology have been authorized and compiled by the Ministry of Education, Science and Culture as a series entitled *Japanese Scientific Terms*, published by appropriate organizations.

● ***440*** *(Commission/Committee)*
Monbushō gakujutsu shingikai gakujutsu yogo bunkakai
[Sub-Committee on Scientific Terms, Science Council, Ministry of Education, Science and Culture]
2–2, Kasumigaseki 3–chome,
Chiyoda-ku Phone: 03–581–4211 Ex.
Tokyo, 100 837
Japan
Contact: Mr. Akira Tabohashi
Director
Science Information and University Library Division, Science and International Affairs Bureau, Ministry of Education, Science and Culture

Committee members: 66
Chairperson: Dr. Ken'ichi Inada
Established: 1967 (continuation of the Scientific Terms Council, Ministry of Education 1949)
Vocabularies:
Subject(s): all fields (23)
UDC 0/9
Language(s): Ja; En
Vocabulary standards:
Subject(s): all fields (34)
UDC 0/9
Language(s): Ja; En
Publications: Results of the Sub-Committee on Scientific Terms are published as *Japanese Scientific Terms* by appropriate organizations

● ***441*** *(Commission/Committee)*
Nihon Igakkai Igakuyōgo Iinkai
[The Committee for Medical Terminology]
Yogo-Iinkai
Parent organization: Nihon Igakkai (Japanese Association of Medical Sciences)
Kanda-Surugadai 2-5,
Chiyodaku Phone: (03)291-2121
Tokyo, 101
Japan
Contact: Dr. Ryuta Ito
Committee members: 158
Chairperson: Dr. Hiroshi Kumagai
Vice-chairperson: Dr. Ryuta Ito
Established: 1940
Vocabularies:
Subject(s): biology; medicine; public health
UDC 575; UDC 576; UDC 577; UDC 578; UDC 61
Language(s): Ja; En; La; De
Vocabulary standards:
Subject(s): medicine; biology
UDC 575; UDC 576; UDC 577; UDC 587; UDC 61
Language(s): Ja; En; La; De
Theory: Ja; En; La; De
Publications: Igakuyōgo Jiten [Japan Medical Terminology], 6th ed. Tokyo: Nanzando, 1980

● ***442*** *(Organization)*
Nihon Kagaku Gijutsu Jōhō Sentā
[The Japan Information Center of Science and Technology]
JICST
Parent organization: Science and Technology Agency
5-2, Nagatacho 2-Chome, Chiyoda-ku
C. P. O. Box 1478 Phone: (03) 581-6411
Tokyo, 100
Japan Telegr.: JICSTECH
Staff: 326
Head: Mr. Tojiro Oka, President
Expert(s): Mr. Hirosi Nakai, Director, Technical Coordination and Development Office
Staff categories:
terminologist(s) (theory) (number unknown); documentalist(s)/librarian(s) (number unknown); 140 information specialist(s) (approximately); translator(s) (number unknown)
Founded: 1957
Documentation:
bibliographic data
factographic data
Subject(s): physical sciences; technology
UDC 5/6
Language(s): Ja
Vocabularies:
Subject(s): general science; technology
UDC 5/6
Language(s): Ja
Vocabulary standards:
Subject(s): bibliographic references
UDC 001. 811; UDC 002
Language(s): Ja
Theory: Ja
Training: graduate
Access to training: JICST Retrieval (DATABASE)
Curriculum: lexicographical methods
Publications: Current *Bibliography on Science and Technology*. This is a compilation of Japanese language abstracts from Japanese and foreign sources. It is published in 11 different series as follows:
Chemistry and Chemical Industry (foreign: three times/month, Japanese: monthly);
Civil Engineering and Architecture (semi-monthly);
Earth Sciences, Mining and Metallurgy (semi-monthly);
Electronics and Electrical Engineering (semi-monthly);
Energy (monthly);
Environmental Pollution (monthly); *Management Science and Systems Engineering* (monthly);
Mechanical Engineering (semi-monthly); *Nuclear Engineering* (monthly);
Pure and Applied Physics (semi-monthly);
Foreign Patent News (Chemical Patents) (weekly publication); *Technical Highlights* (monthly);
Collected Papers of Annual Study Meeting for Information on Science and Technology; *JICST Thesaurus*, 1978 (revised annually);
JICST On-going Research Information Series: Guide to the Public Laboratories and Research Organizations in Japan;
Information and Documentation (Joho Kanri) (monthly)
Services: reference; reproduction; translation; referral; SDI; other: On-line information retrieval services
Service charges
User aids: Newsletter; Texts, guides, etc. (irregular)
Operation:
computer-assisted
Data base description:
Name: JICST
Size: 1, 986 457 records
Language(s): Ja
Data processing:
Hardware: HITAC; HITAC M-17O; 2 Mb; 3 Mb; 4 Mb; 6. 4 Gb; 800bpi/1600bpi, 1750KB/sec 2; 1600bpi/6250bpi, 1250KB/sec 4; magnetic disc unit H8589 200 Mb 40; line printer H-8276 1500 line/min 2; Console display H-8092; 2, Card Reader H-8297 1000 card/min 1; Dial up 300 BPS 580; Leased Line 2400 BPS 72; Leased Line (Kanji) 2400 BPS 7
Software: operating system: EDOS/MSO; operation mode: batch; real time; programming language(s): ASSEMBLER; COBOL; PLI; MASP; data base management system/retrieval documentation system: JOIS/JOIS
Access:
via telephone
on-line; off-line
Data available:
with charge

● ***443*** *(Organization)*
Nihon Kikai Gakkai
[The Japan Society of Mechanical Engineers]
JSME
4-9, Yoyogi 2-chome,
Shibuya-ku Phone: (03)379-6781
Tokyo, 151
Japan Telegr.: KIKAIGAKKAI TOKYO
Staff: 23
Head: Prof. Dr. Kenji Chijiiwa
Expert(s): Mr. Isamu Arai
Founded: 1897
Subject(s): mechanical engineering
UDC 62
Language(s): Ja; En

● ***444*** *(Organization)*
Rikagaku Kenkyūsho
[The Institute of Physical and Chemical Research]
Parent organization: The Agency of Science and Technology
2-1, Hirosawa Phone: 0484-62-1111
Wako, 351
Japan Telegr.: Rikagakuinst

Visiting address: 2–1, Hirosawa, Wako-shi, Saitama-ken, Japan
Staff: 3
Head: Dr. Tatsuoki Miyajima
Expert(s): Dr. Hideaki Sugawara
Staff categories:
1 terminologist(s) (theory); 1 linguist(s)
Founded: 1958
Documentation:
bibliographic data
factographic data
Subject(s): microbiology
UDC 576. 8
Language(s): Ja; En
Vocabularies:
Subject(s): microbiology; tissue and cell culture
UDC 576. 8
Language(s): Ja
Vocabulary standards:
Subject(s): microbiology; tissue and cell culture
UDC 576. 8
Language(s): Ja; En
Publications: Sugawara, H. et al. *Development of a national information system of microorganisms* (draft).
Word lists
Services: reference
No service charges
Operation:
computer-assisted
Data base description:
Name: NISLO
Language(s): Ja; En
Terminological data:
explanations; equivalents
Data processing:
Hardware: IBM; FACOM; 3033MP; M180
Software: operating system: MVS; F4; operation mode: real time; programming language(s): PL1; COBOL
Access:
off-line
Data available:
free of charge

(Project(s))
Responsible: Committee of Information on Microorganisms
Title: Jikken-Seibutsu-Jōhō-Shisutemu [A National Information System of Microorganisms]
Project type: institutional; in progress
Principal investigator: Dr. Daizo Ushiba
Collaborators: 20
Scope and/or subject matter: Tissue and cell cultures
Source material: Published papers and books
Publications: Working papers

○ ***445*** *(Organization)*
Indian Standards Institution
ISI — **Phone: 26 60 21**
Manak Bhavan, 9 Bahadur
Shah Zafar Marg, 110002 — Telex: 031–3970 isi new delhi
India — Telegr.: manaksanstha
Head: Dr. A. K. Gupta, Director General
Vocabulary standards:
Subject(s): all fields
UDC 0/9
Language(s): En

○ ***446*** *(Organization)*
Bureau of Ceylon Standards
BCS
53 Dharmapala Mawatha — Phone: 94 22 60 51
Colombo 3
Sri Lanka — Telegr.: pramika
Head: Mr. K. Herath, Director
Vocabulary standards:
Subject(s): all fields
UDC 0/9
Language(s): En

○ ***447*** *(Organization)*
Pakistan Standards Institution
PSI
39 Garden Road, Saddar — Phone: 7 30 88
Karachi-3
Pakistan — Telegr.: peyasai
Head: Mr. S. M. Shahid
Vocabulary standards:
Subject(s): all fields
UDC 0/9
Language(s): En

○ ***448*** *(Organization)*
Institute of Standards and Industrial Research of Iran. Ministry of Industries
ISIRI
P. O. Box 2937 — Phone: 98 21 89 53 05
Tehran
Iran — Telegr.: standinst
Head: Mr. A. Lavasani, Acting Managing Director
Vocabulary standards:
Subject(s): technical fields
UDC 6
Language(s): Fa

○ ***449*** *(Organization)*
Türk Standardlari Enstitüsü
[Turkish Standards Institute]
TSE
Necatibey Cad. 112,
Bakanliklar — Phone: 18 72 40
Ankara — Telex: tse-tr 42047
Turkey — Telegr.: standard
Head: Mr. Ali Armağan, Secretary General
Vocabulary standards:
Subject(s): all fields
UDC 0/9
Language(s): Tr

○ ***450*** *(Organization)*
Central Organization for Standardization and Quality Control. Planning Board
COSQC
Aljadiria
P. O. Box 13032 — Phone: 776 51 80
Baghdad
Iraq — Telegr.: iros baghdad
Vocabulary standards:
Subject(s): all fields
UDC 0/9
Language(s): Ar

○ ***451*** *(Organization)*
Lebanese Standards Institution
LIBNOR
P. O. Box 19–5144
Beirut
Lebanon
Vocabulary standards:
Subject(s): all fields
UDC 0/9
Language(s): Ar

● *452* *(Organization)*
Friedmann, Mr. Jehuda (private individual)
100 Ben-Jehuda Street
P. O. Box 3320 Phone: 03–231952
Tel-Aviv
Israel Telegr.: As above
Visiting address: As above
Staff: 1 (working full-time in conjunction with experts from various countries)
Head: Mr. Jehuda Friedmann
Expert(s): Mr. Jehuda Friedmann
Documentation: factographic data
Subject(s): work tools
UDC 621. 9
Language(s): He; Ar; Da; Sv; Fi; Is; En; Fr; De; Ru; It; Es; Pt
Vocabularies:
Subject(s): work tools
UDC 621. 9
Language(s): He; Ar; Da; Sv; Fi; Is; En; Fr; De; Ru; It; Es; Pt
Publications: FRIEDMANN, J. *English-Hebrew pictorial dictionary of work-tools.* Tel-Aviv: J. Friedmann, 1971, 28 p., A5; FRIEDMANN, J. *Work-Tools; Dictionary in twelve languages.* Tel-Aviv: J. Friedmann, 1977, 80 p., A5. In addition, monolingual work-tool dictionaries published in He.
Services: reference; translation; referral
Service charges

(Project(s))
[work tools]
Principal investigator: Mr. Jehuda Friedmann
Collaborators: Several experts worldwide
Scope and/or subject matter: Translation and publication of work tools dictionaries in many additional languages
Source material: Material available worldwide
Publications: *Work-Tools Dictionary in Portuguese*

○ *453* *(Organization)*
Standards Institution of Israel
SII
42 University Street Phone: 972 3 42 28 11
Tel Aviv, 69977 Telex: 35508 siit il
Israel Telegr.: standardis
Head: Mrs. Miriam Muller, Director General
Vocabulary standards:
Subject(s): all fields
UDC 0/9
Language(s): He

○ *454* *(Organization)*
The Academy of the Hebrew Language
P. O. Box 3449; 7105
Jerusalem; Tel-Aviv
Israel
Staff: 12 (10 full time, 2 part time). In addition to the salaried staff, some 80 members of the Academy and experts serve as advisors in terminological committees on a voluntary basis.
Head: Prof. Z. Ben-Hayyim, president
Expert(s): Prof. E. Eytan, head of the Terminology Board
Staff categories:
2 documentalist(s)/librarian(s)
Founded: 1953 (The Academy is continuing the activities of the Hebrew Language Council founded 1880.)
Activities: Terminological activities carried out - on the national level - in terminological committees appointed by the Academy in accordance with the law of the Supreme Academy for the Hebrew language.
Documentation: bibliographic data
Subject(s): technology; humanities
UDC 3/7
Language(s): He; En; Fr; De
Vocabularies:
Subject(s): technology; humanities
Language(s): He
Publications: A complete listing of dictionaries published so far by the Academy of the Hebrew Language is contained in the English version *List of publications*, Jerusalem: The Academy of the Hebrew Language, 1974; a list (in Hebrew) of all terminological publications, including those which appeared in periodicals, is also available from the Academy
Services: other: Information on Hebrew terminology is provided to organizations and individuals without charge
Data processing:
Software: A special program for the compilation of all terms contained in the publications of the Academy has been in use for several years

○ *455* *(Organization)*
Thai Industrial Standards Institute. Ministry of Industry
TISI
Rama VI Street Phone: 66 2 281 79 97 - 9
Bangkok 4 Telex: 84 375 moithai th (attention tisi)
Thailand Telegr.: thaistand
Head: Mr. Manus Sooksmarn, Secretary General
Vocabulary standards:
Subject(s): all fields
UDC 0/9
Language(s): En

○ *456* *(Organization)*
Singapore Institute of Standards and Industrial Research
SISIR
179, River Valley Road
Maxwell P. O. Box 2611 Phone: 65 3 36 09 33
Singapore 9046 Telex: rs 28499 sisir
Singapore Telegr.: sisir
Head: Dr. Lee Kum Tatt, Chairmann and Chief Executive
Vocabulary standards:
Subject(s): all fields
UDC 0/9
Language(s): En

○ *457* *(Organization)*
Standards and Industrial Research Institute of Malaysia
SIRIM
Lot 19810, Phase 3, Federal Highway
P. O. Box 35, Shah Alam Phone: 36 26 01
Selangor Telex: ma 38672
Malaysia Telegr.: sirimsec sungei renggam
Head: Mr. Abdullah bin Mohd. Yusof, Controller
Vocabulary standards:
Subject(s): all fields
UDC 0/9
Language(s): En

● *458* *(Organization)*
Institut Bourguiba des langues vivantes
[Bourguiba Institute of Modern Languages]
IBLV
Parent organization: Tunesian National Ministry of Higher Education and Scientific Research
47, Avenue de la Liberté Phone: 282–418; 282–923
Tunis, 1002 Telex: IBLV-TUNIS
Tunesia
Head: Dr. Mohamed Maamouri
Founded: 1963
Documentation:
bibliographic data
factographic data

Subject(s): all fields
UDC 0/9
Language(s): Ar; Fr; En
Theory: Ar; Fr; En
Training: undergraduate; graduate
Access to training: Tunesian university students only
Curriculum: fundamentals of terminology; terminological principles; lexicographical methods. Part of "Maîtrise de Traduction" (first language: Arabic)
Publications: *Language and Linguistics Series*. General Editor: Dr. Mohamed Maamouri, Director of the Institute. a) *English for Special Purposes*; b) *Terminologie Franco-Arabe*. Publication of an Arabic version of ISO 1087 (Vocabulary of terminology) is planned for 1983
Services: reference; other: Exchange of materials
Operation:
manual

(Project(s))
Title: Terminologie de langue Arabe de la phonétique et phonologie [Arabic terminology of phonetics and phonology]
Project type: institutional; in progress
Principal investigator: Dr. M. Maamouri
Scope and/or subject matter: Phonology and phonetics in Arabic; collaboration with the Centre international de recherche et des études en linguistique fondamen tale et appliquée
Method of work: Analysis of Arabic sources
Publications: Envisaged for 1983

○ ***459*** *(Organization)*
Institut National de la normalisation et de la propriété industrielle
Garde Courrier
Tunis-Belvédère, 1002
Tunesia
Vocabulary standards:
Subject(s): all fields
UDC 0/9
Language(s): Fr; Ar

○ ***460*** *(Organization)*
Egyptian Organization for Standardization
EOS
2 Latin America Street,
Garden City — Phone: 29 72 0
Cairo — Telex: 93296 eos un
Egypt, Arab Rep. of — Telegr.: tawhid
Head: Dr. F. A. Sobhy, President
Vocabulary standards:
Subject(s): all fields
UDC 0/9
Language(s): Ar

● ***461*** *(Organization)*

[Bureau of Coordination of Arabization in the Arab World; Bureau de coordination de l'arabisation dans le monde arabe]
Alecso
Parent organization: Arab League Educational, Cultural and Scientific Organization (Alecso)
10, Rue Angola
P. O. Box 290 — Phone: 329–40/326–25
Rabat — Telex: 31851M
Morocco — Telegr.: Tansiktarib
Staff: 50
Head: Prof. Abdelaziz Benabdellah
Expert(s): Dr. Ali M. Al-Kasimi
Staff categories:
3 terminologist(s) (special field); 2 terminologist(s) (theory); 3 documentalist(s)/librarian(s); 4 linguist(s); 2 information specialist(s); 1 translator(s); 2 terminology teacher(s)
Founded: 1961
Documentation: bibliographic data
Subject(s): all fields
UDC 0/9
Language(s): Ar; En; Fr
Operation:
manual

○ ***462*** *(Organization)*
Service de normalisation industrielle marocaine. Direction de l'industrie. Ministère du commerce et de l'industrie
SNIMA
5, rue Arrich — Phone: 201 96
Rabat — Telex: 246 56 m
Morocco
Head: Mr. Amraoui, Directeur
Vocabulary standards:
Subject(s): all fields
UDC 0/9
Language(s): Fr

○ ***463*** *(Organization)*
Institut algérien de normalisation et de propriété industrielle
INAPI
5, rue Abou Hamou Moussa - Centre de Tri
B. P. 1021 — Phone: 213 63 51 80
Alger — Telex: 53086 inapi dz
Algeria — Telegr.: inapi-alger
Head: M. D. Hadj-Sadok, Directeur général
Vocabulary standards:
Subject(s): all fields
UDC 0/9
Language(s): Fr; Ar

○ ***464*** *(Organization)*
Nigerian Standards Organisation. Federal Ministry of Industries
NSO
No. 4 Club Road
P. M. B. 01323 — Phone: 042 33 70 49
Enugu — Telex: 51 108 phoeni ng (attention nso)
Nigeria
Head: Mr. D. O. Ogun, Director
Vocabulary standards:
Subject(s): all fields
UDC 0/9
Language(s): En

● ***465*** *(Organization)*
Institut de recherche d'informatique juridique
Faculté de Droit
P. O. Box 3968
Libreville
Gabon
Staff: 2 collaborators
Head: Prof. Jean-Paul Buffelan
Founded: 1969
Documentation:
bibliographic data
factographic data
Subject(s): law
UDC 34
Training: Training possibilities exist for 20 to 30 persons for legal documentation.
Curriculum: terminological principles
Services: reference; loan; reproduction; other: Services also provided for public administration and the courts
Data processing:
Hardware: IRIS; SO CVII 44K

Software: operating system: DARIUS; programming language(s): Jurindex; Legisdata

(Project(s))
Title: DARIUS
Project type: institutional
Principal investigator: M. P. Rivière
Scope and/or subject matter: Legal information and documentation
Publications: *Le système DARIUS et ses applications à l'informatique juridique* [The DARIUS system and its application in legal data processing]. In: Automatisme, no. 5, 1971.

Title: Jurindex
Project type: institutional
Principal investigator: M. J. -P. Buffelan
Scope and/or subject matter: Analysis of legal documents and presentation in separate alphabetical indexes and chronological tables.
Publications: *Les progrès de l'informatique juridique en France* [The progress of legal data processing in France]. In: Système, no. 1, 1972 (published in Turin, Italy)

Title: Tryptique "Ajuga", "Aladoc" et "Alegi"
Project type: institutional
Principal investigator: M. J. -P. Buffelan
Scope and/or subject matter: Automatization of jurisprudence, documentation and legislation in Gabon
Publications: *Actes de la IX Conférence de la paix mondiale par le droit*, Madrid, 1979

○ ***466*** *(Organization)*
Kenya Bureau of Standards
KEBS
Off Mombasa Road, Behind Belle Vue Cinema
P. O. Box 54974 Phone: 54 12 06/07
Nairobi
Kenya Telegr.: kenstand
Head: Mr. F. B. Maiko, Director
Vocabulary standards:
Subject(s): all fields
UDC 0/9
Language(s): En

● ***467*** *(Organization)*
Taasisi ya Uchunguzi wa Kiswahili
[Institute of Kiswahili Research]
TUKI; IKR
Parent organization: University of Dar Es Salaam
University Hill
P. O. Box 35110 Phone: 53 565
Dar Es Salaam
Tanzania
Staff: 44
Head: Mr. C. W. Temu 6431
Expert(s): Prof. Rajmund Ohly; Mr. Hamisi Akida
Staff categories:
2 terminologist(s) (special field); 2 terminologist(s) (theory); 1 documentalist(s)/librarian(s); 6 linguist (s); 4 translator(s)
Founded: 1964 (formerly Inter-Territorial Language/Swahili Committee 1930 - 1964)
Documentation:
bibliographic data
factographic data
Subject(s): all fields
UDC 0/9
Language(s): Sw; En
Vocabularies:
Subject(s): economics; biology; medicine; physics; civil engineering; masonry; chemistry
UDC 33; UDC 5; UDC 61; UDC 66; UDC 69
Language(s): Sw; En
Vocabulary standards:
Subject(s): economics; geography; technology; domestic science; mathematics; sports
UDC 33; UDC 51; UDC 62; UDC 79; UDC 9
Language(s): En; Sw
Publications: *Kamusi za kiswahili sanifu* (Standard Swahili dictionary)
Services: reference; translation
Service charges
Operation:
manual
Terminological data:
definitions; standardized terms; equivalents

(Project(s))
Title: English/Swahili Dictionary
Project type: institutional; in progress
Principal investigator: Prof. Rajmund Ohly
Collaborators: 5
Scope and/or subject matter: General
Source material: Other dictionaries (English/Kiswahili by Johnson, F.; Snoxal.); word lists, informants, neologisms
Publications: Working papers

Title: Kamusi za masomo maalumu [Dictionaries for special fields]
Project type: institutional; in progress
Principal investigator: Mr. Hamisi Akida
Scope and/or subject matter: Biology; civil engineering; chemistry and physics; medicine
Source material: Field dictionaries; word-lists; informants; neologisms
Publications: Working papers

Title: Terminological Data Bank 612 A
Principal investigator: Ms. Zubeida N. Z. Tumbo
Scope and/or subject matter: Commercial economics and already standardized terms from National Swahili Council
Method of work: Classification of material and card file preparation
Source material: Word-lists from National Swahili Council
Publications: *Glossary for Commercial Economics* (classified) - forthcoming; Working papers

● ***468*** *(Organization)*
Die Suid-Afrikaanse Vertalerinstituut/The South African Institute of Translators and Interpreters
SAIVERT; SAITINT
P. O. Box 3593
Pretoria, 0001
South Africa, Rep. of
Staff: Dr. J. H. Picard, SM, MA SAIT(Chairman); no permanent staff; all language services and many universities employ terminologists/translators some of whom of SAITINT
Expert(s): Dr. J. H. Picard, SM, MA SAIT; Prof. A. D. de V. Cluver; Dr. A. J. Smit
Founded: 1956
Documentation:
bibliographic data
factographic data
Subject(s): all fields
UDC 0/9
Language(s): Af; En; De; Fr; Pt; African languages
Vocabularies: Every institution prepares its own terminology and hands over its research material fo the Terminology Section of the State Language Services, Department of National Education by which it is published.
Subject(s): all fields
UDC 0/9
Language(s): Af; En; Fr; Ja
Training: undergraduate; graduate; postgraduate
Curriculum: Material taught within the framework of linguistics courses of the university.

Publications: A quarterly bulletin, namely: *Die Taalpraktisyn/The Language Practitioner*
Services: reference; translation (supply of information only)

● *469* *(Commission/Committee)*
Terminologievereniging van Suid-Afrika/Terminology Association of South Africa
TSA
Parent organization: Die Suid-Afrikaanse Vertalersinstituut/The South African Institute of Translators and Interpreters
Contact: Mr. P. J. Taljaard, Head, Language Service, Dept. Posts and Telecommunication, c/o General Post Office, Church Street, Pretoria, South Africa; or: c/o Language Bureau of the City Council of Pretoria, P. O. Box 408, Pretoria
Chairperson: Mr. P. J. Taljaard; Ms. E. Viljoen
Vocabularies:
Subject(s): all fields
UDC 0/9
Language(s): Af; En

○ *470* *(Organization)*
South African Bureau of Standards
SABS
Private Bag X191 — Phone: 27 12 48 13 11
Pretoria, 0001 — Telex: 3 626 sa
South Africa, Rep. of — Telegr.: comparator
Head: Mr. G. P. Verster, Director General
Vocabulary standards:
Subject(s): all fields
UDC 0/9
Language(s): En; Af

● *471* *(Organization)*
South African Water Information Centre
[Suid-Afrikaanse Inligtinsentrum vir Water (Nl)]
SAWIC; SAIW (Nl)
Parent organization: Council for Scientific and Industrial Research
P. O. Box 395 — Phone: 86 92 11
Pretoria — Telex: 3630
South Africa — Telegr.: NAVORS
Visiting address: Meiring Naudeweg, Lynnwood, Pretoria
Staff: 13
Head: Mr. M. R. Steyn
Expert(s): Mrs. C. Coetzer
Staff categories:
3 information specialist(s); 10 other professional(s)
Founded: 1974
Subject(s): water
UDC 628
Language(s): En; Af; Nl; De; Fr; Flemish
Vocabularies:
Subject(s): water
UDC 628
Language(s): En; Af; Nl; De; Fr
Publications: *S. A. Waterabstracts* (annually); selected journals on water; *The water industry in South Africa*
Services: reference; referral; SDI; other: Information services
No service charges
Operation:
computerized
Data base description:
Name: Waterlit
Size: 55 000 items
Language(s): En
Classification: Waterlit thesaurus
Data processing:
Hardware: IBM; 370; printer
Software:
operating system: STAIRS; operation mode: real time; programming language(s): PL1
Access:
via telephone/via telex
Data available:
with charge (via Orbit)

● *472* *(Organization)*
Terminologieseksie van die Taaldiensburo van die Department Nasionale Opvoeding
[Terminology Section of the Language Services Bureau of the Department of National Education]
Van der Stel Bldg., Pretorius Street
Postbus X196
Pretoria
South Africa
Staff:
Head: Mr. Laubscher; Mr. Schutte
Documentation:
bibliographic data
factographic data
Subject(s): science; technology; humanities; theoretical linguistics; Afrikaans linguistics; lexicography; information science; communications theory
UDC 0/9
Language(s): Af; En; Fr
Vocabularies:
Subject(s): all fields
UDC 0/9
Language(s): Af; En
Training: Possibilities for training concerning theory and practice of terminology exist
Publications: CLUVER, A. D. de V. *Taalsisteem en die Vaktaalburo.* In: Nuusbrief S. A. *Akademie* 9, 2, 1971, p. 12–16;
CLUVER, A. D. de V. *Drogery-Dwelmi en die Afrikaanse taalsisteem.* In: *Tegnikon 21, 1, 1972, p. 28–32;*
CLUVER, A. D. de V. *Die mens voor die woordeboek versus die mens agter die woordeboek.* In: *Tydskrif vir Geesteswetenskappe* 14, 2, 1974, p. 100–122; VAKTAALBURO. *Terminologierubriek.* In: *The Language Practitioner*, 1974;
In addition, dictionaries for the following fields are available from the Vaktaalburo, namely for: construction, pharmacy, chemistry, survey paper industry, botany, plastics, explosives, statistics, nuclear energy, astronomy, insurance, water and sewage works, mathematics, library, music, etc.
Services: reference; reproduction; other: Answering of terminological queries by mail or phone
No service charges (for information services)

● *473* *(Commission/Committee)*
Comité de linguistique
[Linguistic Committee]
Parent organization: Canadian Broadcasting Corporation
6000 — Phone: 285–2727
Montréal, Québec, H3C 3A8 — Telex: 055–61247
Canada
Visiting address: 1055, Dorchester Blvd East, Montréal, Québec, Canada
Contact: Robert Dubuc
Service de linguistique et de traduction
Société Radio Canada
Committee members: 8
Chairperson: M. Fernand Guérard
Vice-chairperson: M. Robert Dubuc
Established: 1960
Vocabularies:
Subject(s): data processing; management; photography; cinema; audiovisual media; theater; television
UDC 654. 1; UDC 681. 3; UDC 792; UDC 026. 6
Language(s): Fr; En
Publications: Terminological index cards and linguistic bulletins (periodical publications); Pollet, R. J. *Lexique de la photographie d'amateur (vocabulaire et définitions bilingues des termes employés en photographie d'amateur).* 111 p., 3, 75 $; *Lexique de la cinématographie d'amateur* (vocabulaire et définitions bilingues des termes employés en cinématographie d'amateur), 128 p., 4, 00 $. *Lexique des termes techniques (liste de 1600 termes anglais- français*

touchant les principaux aspects de l'audiovisuel) 233 p., 9, 50 $; Dubuc, R. Objectif 200 (Deux cent fautes de langages à corriger), 133 p., 3, 50$; *Vocabulaire de gestion (vocabulaire bilingue avec définitions des principaux termes utilisés en gestion),* 137 p., 4, 50 $; *Vocabulaire bilingue du théâtre* (près de 1300 termes étudiés en anglais et en français pour concilier deux univers théâtraux en constante interaction), 174 p., 7, 95 $

(Project(s))
Title: Vocabulary of Statistical Research (bilingual En-Fr), (deferred)
Scope and/or subject matter: statistics
Method of work: The vocabulary will comprise 500 entries.

Title: Vocabulary of Television Production (bilingual Fr-En)
Scope and/or subject matter: Television production
Method of work: The vocabulary will comprise 1200 entries.

● ***474*** *(Organization)*
Department of French, Faculty of Arts and Science, Dalhousie University
[Département de français, Faculté des arts et de la science, Université Dalhousie] Phone: 1–902–424–2430
Halifax, Nova Scotia, B3H 3J5
Canada
Head: Prof. Dr. Roseann Runte
Expert(s): Prof. Dr. Rostislav Kocourek
Staff categories:
11 linguist(s); 13 other professional(s)
Founded: 1818
Theory: Fr; En
Training: undergraduate; graduate
Publications: R. Kocourek's writings in specialized language and terminology (appr.40 items)

(Project(s))
Title: La langue française de la technique et de la science [The French language in technology and science]
Project type: commissioned project: (to be published by Oscar Brandstetter Verlag KG, Wiesbaden); in progress
Principal investigator: Rostislav Kocourek
Collaborators: 1
Scope and/or subject matter: Linguistic characteristics of technical and scientific French with emphasis on terminology (ch.1: delimitation and differentiation, ch.2: textual, syntactic, lexical and graphical specificity, ch.3 to 4: formation, meaning and structure of terminology, ch.5: applications, perspectives, evaluation).
Method of work: Analysis of primary data and synthesis
Source material: Contemporary technical and scientific texts (including glossaries); technical-linguistic, lexicological and terminological studies.
Publications: *Prerequisites for an applicable linguistic theory of terminology.* In: Savard, J. -G.; Laforge, L. (eds.) *Proceedings of the 5th International Congress of AILA,* Montréal, Aug. 1978. Québec: Presses de l'Universite Laval, 1981, p. 216–218;
Lexical phrases in terminology. Travaux de terminologie, vol. 1, no. 1, oct. 1979, p. 121–153; *La nature du français technique et scientifique,* Wiesbaden: O. Brandstetter, 1980, vol. 4, p. 1153–1174.

● ***475*** *(Organization)*
Direction Générale de la terminologie et de la documentation/Terminology and Documentation Branch
DGTD/TDB
Parent organization: Bureau des traductions, Secrétariat d'État, Gouvernement du Canada/Translation Bureau, Secretary of State, Canadian Government
15, Eddy Street Phone: (819) 997–1321
Ottawa, K1A OM5 Telex: 0533384
Canada
Visiting address: 7th floor, 15 Eddy Street, Hull, Québec, Canada
Staff: 217
Head: Mr. Alain Landry, Director General; Ms. S. Richer, Acting Director General
Expert(s): Mr. Michel Cardin, Director
Staff categories:
5 terminologist(s) (special field); 130 terminologist(s) (theory); 20 documentalist(s)/librarian(s); 4 linguist (s); 14 translator(s); 8 terminology teacher(s); 36 other professional(s)
Founded: 1975 (DGTD); 1934 (Translation Bureau)
Documentation:
bibliographic data
factographic data
Subject(s): all subjects
UDC 0/9
Language(s): Fr; En; Da; Sv; No; De; Es; Ru; Ja; It;
Vocabularies:
Subject(s): all fields
UDC 0/9
Language(s): Fr; En
Vocabulary standards:
Subject(s): all fields
UDC 0/9
Language(s): Fr; En
Theory: Fr; En
Training: undergraduate
Access to training: available to outside users
Curriculum: fundamentals of terminology; terminological principles
Publications: *Bulletin de terminologie/Terminology Bulletin* (monographs published irregularly);
Actualité terminologique (Monthly bulletin of the DGTD).
The following research projects have been finished in the last couple of years and either lists of equivalents or data on card files are available on the following subject fields:
Additifs alimentaires/food additives; pesticides; energy efficient buildings; réfection des toitures/repairing of roofs; appel d'offres et marchés/public tender; lasers; traitement des eaux usées/waste-water treatment; gestion des documents/administration of documents; criblage industriel/industrial screening; couture industrielle/ready-to-wear industry; métallurgie - essais non destructifs des métaux/metallurgy - non-destructive testing of metals; métallurgie - essais mécaniques des métaux/metallurgy - mechanic testing of metals; industrie alimentaire/food industry; visserie - boulonnerie/screwing - bolting; équipement du cabinet dentaire/dental equipment; maladies cérébro-vasculaires/cerebral-vascular diseases; mé decine nucléaire/nuclear medicine; occupational therapy; lutte contre la pollution par les hydrocarbures/fight against the pollution through hydrocarbons; post office; champignons/mushrooms; Quebec educational system; radar
Publications are listed selectively in *Terminology Bulletin*
Services: reference; loan; reproduction; translation; referral; Being planned; other: Terminology Services
User aids: Bulletins de terminologie/Terminology Bulletins; Actualité terminologique; Répertoire des travaux de recherche/Directory of research projects
Operation:
computer-assisted
computerized
Data base description:
Name: TERMIUM
Size: 6400 RB
Language(s): Fr; En
Classification: Thesaurus (in preparation); in-house system
Terminological data:
definitions; explanations; classified layout; context; equivalents
Data processing:
Hardware: CYBER; 74; 2G2K (1 Main frames); 640 Mb (b 6 bits); BACK UP; DATAPOINT 1100 (printer); DATAPOINT 6600 (data input); 80 Terminals (digital)
Software: operating system: NOS/BE; operation mode:

batch; real time; programming language(s): Fortran; Compass
Access:
via telephone
on-line; off-line
Data available:
on exchange basis; free of charge

(Project(s))
Responsible: Comité d'uniformisation de la terminologie aéronautique [Committee for the standardization of aeronautic terminology]
Title: Aéronautique [Aeronautics]
Project type: institutional; in progress
Principal investigator: Lise Boisvert, scientific and parliamentary section
Scope and/or subject matter: Basic terminology for air operations
Method of work: The terminological research methods adopted by the Translation Bureau, in some cases complemented by a committee of experts
Publications: List of equivalents

Title: Affaires extérieures [Foreign affairs]
Project type: institutional; in progress
Principal investigator: Francine Persaud, economic and legal section
Scope and/or subject matter: International law and diplomacy
Method of work: The terminological research methods adopted by the Translation Bureau, in some cases complemented by a committee of experts
Source material: 30, mostly specialized, works

Title: Chimie [Chemistry]
Project type: institutional; in progress
Principal investigator: Dorothy Nakos-Aupetit, subsection of Quebec
Scope and/or subject matter: Chemical products
Method of work: The terminological research methods adopted by the Translation Bureau, in some cases complemented by a committee of experts
Source material: 10 specialized works
Publications: List of equivalents

Title: Classification des postes [Classification of lots]
Project type: institutional; in progress
Principal investigator: Yolande Labelle, socio-administrative section
Scope and/or subject matter: Terminology used by the Treasury Council in the field of classification of lots
Method of work: The work will be presented to a committee of experts
Source material: 43, mostly specialized, works

Responsible: Study Group on Accounting Terminology/Groupe d'étude de la terminologie comptable (Comité mixte - secteurs privé et public)
Title: Comptabilité publique [Public accounting]
Project type: institutional; in progress
Principal investigator: Maurice Perrier, socio-administrative and economic section
Scope and/or subject matter: Accounting terms of the federal administration
Method of work: The terminological research methods adopted by the Translation Bureau, in some cases complemented by a committee of experts
Source material: 50, mostly specialized, works
Publications: Publication of an article in *l'Actualité Terminologique* (November 1979, vol.12, no.9) and of parts of the vocabulary studied by the Study Group

Title: Détection et identification des nappes de pétrole [Detection and identification of oil slicks]
Project type: institutional; in progress
Principal investigator: Hélène Jacob, section of Montréal
Method of work: The terminological research methods adopted by the Translation Bureau, in some cases complemented by a committee of experts
Source material: 10 specialized works

Title: DOBIS
Project type: institutional; in progress
Principal investigator: Rita Bélanger, socio-administrative section
Scope and/or subject matter: Bibliotheconomy; informatics; diverse users guide
Method of work: The consultants are the librarians of the Canadian National Library (BNC) and of universities as well as system analysts and the office for bibliographic standards of the BNC
Source material: 50 specialized works
Publications: List of equivalents with context

Title: Dotation en personnel [Staff dotation]
Project type: institutional; in progress
Principal investigator: Yolande Labelle (for the French part); Mary Sitarski (for the English part)
Scope and/or subject matter: Terminology used by the Commission of Public Function in the field of staff dotation
Method of work: The terminological research methods adopted by the Translation Bureau, in some cases complemented by a committee of experts
Source material: 155, mostly specialized, works

Title: Droit des transports maritimes [Law of naval transports]
Project type: institutional; in progress
Principal investigator: Jean Moreault, section of Montréal
Scope and/or subject matter: Terminology of international conventions on naval transport; accidents at sea; contract conditions of marine insurance; regulations of the International Chamber of Commerce and of naval transport conventions submitted to the Federal Court of Canada
Method of work: The terminological research methods adopted by the Translation Bureau, in some cases complemented by a committee of experts
Source material: 40, mostly specialized, works
Publications: An article will be published in *l'Actualité Terminologique*; keeping of files on matters of dispute and, eventually, standardization

Title: Droit maritime international [International maritime law]
Project type: institutional; in progress
Principal investigator: Jean Moreault, Section of Montréal
Scope and/or subject matter: Arrest of seagoing ships
Method of work: The terminological research methods adopted by the Translation Bureau, in some cases complemented by a committee of experts
Source material: 3 specialized works

Title: Effets écologiques d'un déversement de pétrole [Ecological effects of a diversion of oil]
Project type: institutional; in progress
Principal investigator: Madeleine Brisebois, section of Montréal
Scope and/or subject matter: Pollution of shores and estuaries
Method of work: The terminological research methods adopted by the Translation Bureau, in some cases complemented by a committee of experts
Source material: 10 specialized works

Responsible: Comité de normalisation de la terminologie nucléaire (Comité intergouvernemental) [Committee on the standardization of nuclear terminology (intergovernmental committee]

Title: Energie nucléaire [Nuclear energy]
Project type: institutional; in progress
Principal investigator: André Lortie, scientific and parliamentary section
Scope and/or subject matter: Pollution through nuclear waste; techniques of treatment; conditioning and eliminating of waste; regulations in the field of nuclear energy
Method of work: The terminological research methods adopted by the Translation Bureau, in some cases complemented by a committee of experts
Source material: 10, mostly specialized, works
Publications: List of equivalents

Title: Froid [Refrigeration]
Project type: institutional; in progress
Principal investigator: Michèle Valiquette, section of Montréal
Scope and/or subject matter: Basic concepts and measuring instruments; thermodynamics; refrigeration; compressors; exchangers; installations
Method of work: The terminological research methods adopted by the Translation Bureau, in some cases complemented by a committee of experts
Source material: 8 general and 7 specialized works

Title: Géotechnique [Geotechnics]
Project type: institutional; in progress
Principal investigator: Claude Tremblay, economic and legal section
Scope and/or subject matter: Soil mechanics; rock mechanics; civil engineering
Method of work: One part of this work consists of reclassifying 5 specialized lexicons
Source material: Specialized works (120 monographies, 175 articles, 15 dictionaries)
Publications: List of equivalents

Title: Gestion financière [Financial administration]
Project type: institutional; in progress
Principal investigator: Marielle Hébert-Charette, section of Montréal
Scope and/or subject matter: Financial administration; accounting; national economy
Method of work: Bilingual extracts of manuals of the Treasury Council
Source material: 100, mostly specialized, works

Title: High-Fidelity
Project type: institutional; in progress
Principal investigator: Patrick McNamer, English section
Method of work: The terminological research methods adopted by the Translation Bureau, in some cases complemented by a committee of experts
Source material: 76 works, primarily specialized, including a number of standards
Publications: The contents of the glossary are being processed for storage in the bank and preparations are underway to have the vocabulary printed for distribution among Bureau translators

Title: Ichtyologie [Ichthyology]
Project type: institutional; in progress
Principal investigator: Dorothy Nakos-Aupetit, sub-section of Quebec
Method of work: The terminological research methods adopted by the Translation Bureau, in some cases complemented by a committee of experts
Source material: 15 general and specialized works
Publications: List of equivalents

Title: Immigration
Project type: institutional; in progress
Principal investigator: Silvia Pavel, socio-administrative section
Scope and/or subject matter: Legislation; ministerial guides; fichiers
Method of work: The terminological research methods adopted by the Translation Bureau, in some cases complemented by a committee of experts
Source material: 12 specialized works
Publications: List of equivalents

Title: Industrial machinery
Project type: institutional; in progress
Principal investigator: Patrick McNamer, English section
Scope and/or subject matter: Tool room machinery as well as machinery accessories and parts
Method of work: The terminological research methods adopted by the Translation Bureau, in some cases complemented by a committee of experts
Source material: The initial list of terms for this project was derived from a text for translation submitted for research of the terminology

Responsible: Comité des normes gouvernementales en informatique - CNGI (Comité interministériel) [Committee for governmental standards on informatics - CNGI interministerial committee]
Title: Informatique [Informatics]
Project type: institutional; in progress
Principal investigator: Noella Bordage, socio-administrative section
Scope and/or subject matter: Arithmetical and logical operations, data organization, programming, etc.
Method of work: The terminological research methods adopted by the Translation Bureau, in some cases complemented by a committee of experts
Source material: 30, mostly specialized, works

Title: Juridiction [Jurisdiction]
Project type: institutional; in progress
Principal investigator: Claude Tremblay, economic and legal section
Scope and/or subject matter: Different sub-fields of jurisdiction
Method of work: The terminological research methods adopted by the Translation Bureau, in some cases complemented by a committee of experts

Title: Le bois, source d'énergie [Wood, an energy source]
Project type: institutional; in progress
Principal investigator: Dorothy Nakos-Aupetit, sub-section of Québec
Method of work: The terminological research methods adopted by the Translation Bureau, in some cases complemented by a committee of experts
Source material: 15 general and specialized works

Title: Les revenus de retraite au Canada [Retirement revenues in Canada]
Project type: institutional; in progress
Principal investigator: Madeleine Brisebois, section of Montréal
Scope and/or subject matter: Retirement revenues; economy of the social security; demography (population analysis); accounting
Method of work: The terminological research methods adopted by the Translation Bureau, in some cases complemented by a committee of experts
Source material: 12, mostly specialized, works
Publications: List of equivalents

Title: Loi sur les Indiens [Law for the Indians]
Project type: institutional; in progress
Principal investigator: Jean Moreault, section of Montréal
Scope and/or subject matter: Administrative law; historical treaties between the Crown and the Indian tribes (evolution from 1783 until 1956); administration of territories resigned or reserved for the tribes
Method of work: The terminological research methods adopted by the Translation Bureau, in some cases complemented by a committee of experts
Source material: 17 specialized works
Publications: The terminologists will produce a bilingual dictionary of fundamental (historical) terms, which will be distributed to ministries co-operating with the translation section

Title: Lutte contre la pollution par les hydrocarbures [Fight against the pollution through hydrocarbons]
Project type: institutional; in progress
Principal investigator: Denis Rivard, section of Montréal
Scope and/or subject matter: Mechanical means of fighting
Method of work: The terminological research methods adopted by the Translation Bureau, in some cases complemented by a committee of experts
Source material: 25 specialized works

Title: Matériel lourd [Heavy material]
Project type: institutional; in progress
Principal investigator: André Lortie, scientific and parliamentary section
Scope and/or subject matter: Lorries; farming tractors; disc harrows; heavy trucks; construction machinery: wheel tractors and caterpillar tractors, hydraulic dredging machines, scrappers, motor graders
Method of work: The terminological research methods adopted by the Translation Bureau, in some cases complemented by a committee of experts
Source material: 10 specialized works
Publications: List of equivalents

Title: Mechanics: standard commodity classification numeric list
Project type: institutional; in progress
Principal investigator: Dorothy Nakos-Aupetit, sub-section of Québec
Method of work: The terminological research methods adopted by the Translation Bureau, in some cases complemented by a committee of experts
Source material: 15 general and specialized works
Publications: List of equivalents

Title: Métaux [Metals]
Project type: institutional; in progress
Principal investigator: Dorothy Nakos-Aupetit, sub-section of Quebec
Method of work: The terminological research methods adopted by the Translation Bureau, in some cases complemented by a committee of experts
Source material: 15 general and specialized works
Publications: List of equivalents

Title: Mining
Project type: institutional; in progress
Principal investigator: Mary Sitarski, English Section
Scope and/or subject matter: Extraction and mineral processing
Method of work: This project was undertaken in connection with the Summer Job Corps Programme and all the initial research was carried out in Sudbury by students working in contact with a number of mining companies
Source material: 100 works (the majority specialized and monolingual French or English)
Publications: Fully documented terminology records to be entered in the terminology file - also, a two-part glossary produced by the OLF *(Lexique anglais-français de l'industrie minière*: 1.L'exploitation et 2.Le traitement des minérais) is being entered into the bank as a bulk-loading project to supplement the records

Title: Plateformes de forage en mer [Drilling platforms at sea]
Project type: institutional; in progress
Principal investigator: Michèle Valiquette, section of Montréal
Scope and/or subject matter: Platform types and drilling techniques at sea
Method of work: The terminological research methods adopted by the Translation Bureau, in some cases complemented by a committee of experts
Source material: 20 general and specialized works

Title: Poids et mesures [Weights and measures]
Project type: institutional; in progress
Principal investigator: Dorothy Nakos-Aupetit, sub-section of Québec
Method of work: The terminological research methods adopted by the Translation Bureau, in some cases complemented by a committee of experts
Source material: 10 general works
Publications: List of equivalents

Responsible: Comité de normalisation de la terminologie du lexique laitier (participation de l'Office de la langue française) [Committee for the standardization of the terminology of the dairy lexicon (participation of the OLF)]
Title: Produits laitiers [Dairy products]
Project type: institutional; in progress
Principal investigator: André Lortie and Marie-Hélène Borsanyi, scientific and parliamentary section
Scope and/or subject matter: Milk; butter; cheese; yogurt; icecream; etc.
Method of work: The terminological research methods adopted by the Translation Bureau, in some cases complemented by a committee of experts
Source material: The work is carried out on the basis of the Lexique laitier d'Agriculture du Canada
Publications: List of equivalents with contexts or definitions

Title: Propriété foncière - réserve des Indiens [Real property - native reserve of the Indians]
Project type: institutional; in progress
Principal investigator: Jean Moreault, section of Montréal
Scope and/or subject matter: Civil law; proprietary rights concerning the Indians in Canada; the patrimony of their respective tribe; real property and personal property; real servitudes; administration of reserved territories; etc.
Method of work: Setting up of French equivalents of a glossary of 50 English terms with definitions provided by the client
Source material: 5 specialized works

Title: Radioprotection
Project type: institutional; in progress
Principal investigator: André Lortie, scientific section
Scope and/or subject matter: Nuclear medicine; effects of radiations; security of nuclear installations
Method of work: The terminological research methods adopted by the Translation Bureau, in some cases complemented by a committee of experts
Source material: 10 specialized works
Publications: Summary of reports of the World Health Organization (WHO), possibly followed by terminological research; List of equivalents

Title: Rémunération [Remuneration]
Project type: institutional; in progress
Principal investigator: Laurent Buendia, socio-administrative section
Scope and/or subject matter: Terminology used by the Treasury Council in the field of remuneration
Method of work: This work will be studied by a committee of experts
Source material: 36, mostly specialized, works

Title: Road construction
Project type: institutional; in progress
Principal investigator: David Martin, English section
Scope and/or subject matter: Geodetic surveying; cartography; soil testing and materials
Method of work: The terminological research methods adopted by the Translation Bureau, in some cases complemented by a committee of experts
Publications: Terminology records (roughly 35) are being prepared for entry into the bank; List of equivalents

Responsible: Comité de normalisation de la terminologie des services sociaux (Comité mixte - secteurs privé et public) [Committee on standardization of terminology of social services (Mixed committee - private and public sections)]
Title: Services sociaux [Social services]
Project type: institutional; in progress
Principal investigator: For the French part: Casimir Gajda, socio-administrative section; for the English part: Mariam Bugara, English section
Scope and/or subject matter: Social work; education
Method of work: The terminological research methods adopted by the Translation Bureau, in some cases complemented by a committee of experts
Source material: 125 specialized works

Title: Ski de fond [Cross-country skiing]
Project type: institutional; in progress
Principal investigator: Bruno Couture, section of Montréal
Scope and/or subject matter: Climatology; avalanches; camping; shelters; knots; survival; etc. (deals in reality with fields connected to cross-country skiing)
Method of work: The terminological research methods adopted by the Translation Bureau, in some cases complemented by a committee of experts

Title: Soil science
Project type: institutional; in progress
Principal investigator: Mariam Bugara, English section
Scope and/or subject matter: Soil mapping; cartography
Method of work: The terms in this project were derived from a text for translation submitted for research of the terminology
Source material: 32 works, including some specialized sources
Publications: An existing glossary available in Fr and En, the *Glossary of Soil Science*, will be bulk-loaded into the bank along with a small number of records; List of equivalents

Responsible: Comité de spécialistes du Centre national du sport et de la récréation [Committee of specialists of the national centre of sport and recreation]
Title: Sports
Project type: institutional; in progress
Principal investigator: Claude Tremblay, economic and legal section
Scope and/or subject matter: Hockey; swimming; artistic skating; speed skating; etc.
Method of work: The terminological research methods adopted by the Translation Bureau, in some cases complemented by a committee of experts

Title: Statistique [Statistics]
Project type: institutional; in progress
Principal investigator: Louise Lalonde, socio-administrative section
Scope and/or subject matter: Collection and classification; data presentation; data reduction; data analysis; probability; general terms.
Method of work: The terminological research methods adopted by the Translation Bureau, in some cases complemented by a committee of experts
Source material: 165 specialized and 13 general works

Title: Télécommunications [Telecommunications]
Project type: institutional; in progress
Principal investigator: Claude Tremblay, economic and legal section
Scope and/or subject matter: Telephony; satellites; telecommunication instruments
Method of work: Updating of work carried out by students

Title: Télédétection des ressources forestières [Teledetection of wood resources]
Project type: institutional; in progress
Principal investigator: Dorothy Nakos-Aupetit, section of Montréal, sub-section of Québec
Method of work: The terminological research methods adopted by the Translation Bureau, in some cases complemented by a committee of experts
Source material: 15 general and specialized works

Title: Terminology
Project type: institutional; in progress
Principal investigator: Mariam Bugara, English section
Scope and/or subject matter: Primarily terminological research; recording of data; processing of records; terminology bank
Method of work: The terminological research methods adopted by the Translation Bureau, in some cases complemented by a committee of experts
Publications: A tentative base list of the terms specific to terminology as practiced within the Canadian Government is being prepared with a view to publication as an in-house vocabulary

● ***476*** *(Commission/Committee)*
Comité d'uniformisation de la terminologie des marchandises dangereuses
[Dangerous Goods Terminology Standardization Committee (not official)]
Parent organization: Military and Technical Section, Terminology Directorate, Translation Bureau, Secretary of State
Contact: Adeline Lemieux, Terminologist 5th Floor, Centre Block South, 101 Colonel 8y Drive, Ottawa
Committee members: 5
Chairperson: Ms. Adeline Lemieux
Established: 1980
Vocabulary standards:
Subject(s): dangerous goods
UDC 656. 833. 5
Language(s): Fr; En

● ***477*** *(Commission/Committee)*
Comité d'uniformisation de la terminologie aéronautique (contrôle de la circulation aérienne)
[Aeronautical Terminology Standardization Committee (Air Traffic Control) (not official)]
CUTA (ATC)
Parent organization: Civil Aeronautics Branch, Canadian Air Transportation Administration, Transport Canada
Podium Building, Place de Ville
Ottawa, Ontario, K1A ON8
Canada
Contact: Ms. Lise Boisvert, Translation Bureau, Secretary of State Dept., P. O. Box 970, Postal Station "La Cité", Montréal, Quebec, Canada H2W 2R1

Committee members: 9
Chairperson: Ms. Lise Boisvert
Established: 1979

● ***478*** *(Commission/Committee)*
Comité d'uniformisation de la terminologie aéronautique (Opérations aériennes)
[Aeronautic Terminology Standardization Committee (Flight Operations)]
CUTA (Air Canada)
Parent organization: Air Canada, Linguistic Services
5th floor, Computer Building, Air Canada Base
Corval, Québec, H4Y 1C2
Canada
Contact: Ms. Lise Boisvert, Translation Bureau, Secretary of State Dept., P. O. Box 970, Postal Station "La Cité", Montréal, Québec, Canada
Committee members: 11
Chairperson: Ms. Lise Boisvert
Established: 1977
Vocabulary standards:
Subject(s): aeronautics
UDC 629. 7
Language(s): Fr; En

● ***479*** *(Commission/Committee)*
Comité d'uniformisation de la terminologie de l'électrotechnique et des télécommunications
[Committee for the Standardization of Electrotechnical and Telecommunications Terminology (not official)]
Parent organization: Military and Technical Section, Terminology Directorate, Translation Bureau, Secretary of State
Visiting address: 5th Floor, Centre Block South, 101 Colonel By Drive, Ottawa
Contact: Ms. F. Pauline Cormier, Project Officer, Terminology Directorate, Translation Bureau, Secretary of State Dept., Ottawa, Ontario K1A OM5, Canada
Chairperson: Ms. F. Pauline Cormier
Established: 1980

● ***480*** *(Commission/Committee)*
Comité d'uniformisation de la terminologie de la mécanisation postale
[Postal Mechanization Terminology Standardization Committee]
Parent organization: Technical Training Centre, Post Office Department
1495 Heron Road, Pavillon "M"
Phone: 998–9760
Ottawa, Ontario
Canada
Contact: Mr. David Snook, Project Officer, Terminology Directorate, Translation Bureau, Secretary of State Dept., Ottawa, Ontario K1A OM5, Canada
Committee members: 6
Chairperson: Mr. David Snook
Established: 1979
Vocabulary standards:
Subject(s): postal services
UDC 656. 8
Language(s): Fr; En

● ***481*** *(Commission/Committee)*
Comité de normalisation de la terminologie des services sociaux
[Social Services Terminology Standardization Committee]
Parent organization: Terminology Directorate, Translation Bureau, Secretary of State Dept.
Ottawa, Ontario, K1A OM5
Canada
Visiting address: 15 Eddy Street, Hull, Quebec
Contact: Mr. David Snook, Project Officer (at above address)
Committee members: 10
Chairperson: Mr. David Snook
Established: 1979
Vocabulary standards:
Subject(s): social services
UDC 36
Language(s): Fr; En

● ***482*** *(Commission/Committee)*
Comité de normalisation de la terminologie nucléaire
[Nuclear Terminology Standardization Committee (not official)]
Parent organization: Scientific and Parliamentary Section, Terminology Directorate, Translation Bureau, Secretary of State Dept.
Ottawa, Ontario, K1A OM5
Canada
Visiting address: 3rd Floor, 15 Eddy Street, Hull, Quebec
Contact: Mr. David Snook, Project Officer (at above address)
Committee members: 8
Chairperson: Mr. David Snook
Established: 1978
Vocabulary standards:
Subject(s): nuclear energy
UDC 621. 039
Language(s): Fr; En

● ***483*** *(Commission/Committee)*
Comité de terminologie des produits laitiers
[Dairy Products Terminology Committee (not official)]
Parent organization: Secteur de l'alimentation, Direction de la terminologie, Office de la langue française
700, boul. St-Cyrille est Phone: 643–2134
Quéec, Québec, GR7 5A9
Canada
Contact: Mr. Jacques Maurais
Committee members: 8
Chairperson: Mr. Jacques Maurais, Office de la langue française
Established: 1980
Vocabulary standards:
Subject(s): electrical engineering
UDC 621. 3
Language(s): Fr; En

● ***484*** *(Commission/Committee)*
Comité de terminologie du Solliciteur général
[Solicitor General Terminology Committee (not official)]
Parent organization: Solicitor General Section, Legal Group, Departmental Translation Services Directorate, Translation Bureau
Sir Wilfrid Laurier Building
Ottawa, Ontario, K1A OM5
Canada
Visiting address: 340 Laurier Ave. W., 12th Floor, Ottawa, Ontario
Contact: Mr. Jean-Marie Blanchet, Project Officer, Economic and Legal Section, Terminology Directorate, Translation Bureau, Secretary of State Dept., Ottawa, Ontario K1A OM5, Canada
Committee members: 10
Chairperson: Mr. Jean-Marie Blanchet
Established: 1979
Vocabulary standards:
Subject(s): law
UDC 34
Language(s): Fr; En

● ***485*** *(Commission/Committee)*
Comité de vocabulaire du SCC
[SCC Vocabulary Committee]
Parent organization: Standards Council of Canada (SCC), Standards Division
178 Rexdale Boulevard
Rexdale, Ontario, M9W 1R3
Canada
Contact: Ms. Monique Larichellière, Project Officer, Socio-administrative Section, Terminology Directorate, Translation Bureau, Secretary of State Dept., Ottawa,

Ontario, K1A OM5, Canada
Committee members: 14
Chairperson: Mr. Philippe Tessier
Established: 1976
Vocabulary standards:
Subject(s): all fields
UDC 0/9
Language(s): Fr; En

● *486* *(Commission/Committee)*
Government EDP Standards Committee, Task Force on Terminology
[Comité des normes gouvernementales en informatique, Groupe de travail en terminologie]
GESC; CNGI
Parent organization: Supply and Services Canada Advisory Bureau for Computing
Lord Elgin Plaza — Phone: 995-6334
Ottawa, Ontario, K1A OS5
Canada
Visiting address: 66 Slater Street, Ottawa, Ontario
Contact: Ms. Monique Larichellière, Project Officer, Socio-administrative Section, Terminology Directorate, Translation Bureau, Secretary of State Dept., Ottawa, Ontario, K1A OM5, Canada
Committee members: 9
Chairperson: Philippe Tessier
Established: 1978
Vocabulary standards:
Subject(s): data processing
UDC 657. 1
Language(s): Fr En

● *487* *(Commission/Committee)*
Groupe d'étude de la terminologie comptable
[Study Group on Accounting Terminology]
SGAT; GETC
Parent organization: (1) Auditor General of Canada and (2) Supply and Services Canada Accounting Branch
Ottawa, Ontario, (1) K1A OG6; (2) K1A OS5
Canada
Visiting address: (1) 240 Sparks St., Ottawa, Ontario; (2) 11 Laurier St., Hull, Québec
Contact: Mr. Richard Ranger, Office of the Comptroller General of Canada, Treasury Board, Ottawa, Ontario, K1A 1E4, Canada
Committee members: 9; a number of correspondents
Chairperson: Mr. Richard Ranger, Office of the Comptroller-General
Vice-chairperson: Mr. Maurice Perrier, Translation Bureau
Established: 1974
Vocabulary standards:
Subject(s): accounting
UDC 657. 1
Language(s): Fr; En

● *488* *(Data Bank)*
La Banque de terminologie du Gouvernement canadien/Canadian Government Terminology Bank Termium
Parend organization: Bureau des traductions, Secrétariat d'Etat, Gouvernement du Canada/Translation Bureau, Secretary of State, Canadian Government
15 Eddy Street — Phone: (819)997-1321
Ottawa, K1A OM5 — Telex: 0533384
Canada
Visiting address: 7th Floor, 15 Eddy Street, Hull, Québec, Canada
Staff: 217
Head: Ms. Suzanne Richer, Acting Director General
Expert(s): Mr. Michel Cardin
Staff categories:
5 terminologist(s) (special field); 2 terminologist(s) (theory); 6 documentalist(s)/librarian(s) (thesaurus); 8 terminology teacher(s); 10 other professional(s)
Founded: 1975 (DGTD); 1934 (Translation Bureau)
Purpose of data bank/users: Provide terminology services to Federal Government; coordinate standardization of terminologies; co-operate with other terminological organizations; Translators, researchers, writers, Parliament, other public servants
Operation: computer-assisted
computerized
Data base description:
Name: TERMIUM II
Size: 6400 RB
Language(s): Fr; En
Subject(s): all fields
UDC 0/9
Language(s): Fr; En
Terminological data:
definitions; explanations; classified layout; context; equivalents
Products: Publications listed selectively in "Terminology Bulletin"; microforms (production of microfiches is planned); magnetic tapes, disks
(available for data exchange purposes)
Services: Being planned; SVP (telephone information service)
Service charges: Magnetic support, computer time etc. charged at cost
No service charges: Terminological services free of charge
Access:
via telephone
on-line; off-line
Data available:
on exchange basis; free of charge
Data processing:
Hardware/Capacity: CYBER 74; 262K (2 main frames); 640 Mb (b = bb.TS); BACKUP; DATA POINT 1100; DATA POINT 6600; TERMINALS
Software:
operating system: NOS/BE; operation mode: batch; real time; programming language(s): Fortran; Compass

◐ *489* *(Organization)*
Faculty of Library and Information Science. Library, Subject Analysis Systems Collection
140 St. George Street — Phone: (416) 978 7060
Toronto, M5S 1A1
Canada, Ontario
Staff: no dedicated staff
Head: Ms. Diane Henderson, Chief Librarian
Staff categories:
3 documentalist(s)/librarian(s)
Founded: 1976
Documentation: bibliographic data
Subject(s): all fields
UDC 0/9
Language(s): En; Fr; De
Services: reference; loan; reproduction
Service charges
User aids: Catalogue in card form
Operation:
computer-assisted; Computer facilities used are external and at the University of Toronto's computer centre
Data base description:
Name: SAS catalogue
Size: 2 000 records
Language(s): En
Classification: DDC, LCSH; By type of scheme
Access:
via telephone
Data available:
with charge

● *490* *(Organization)*
Groupe interdisciplinaire de recherche scientifique et appliquée en terminologie
[Interdisciplinary Group for Scientific and Applied Terminological Research]
GIRSTERM — **Phone: (418) 565-5492**
Québec, GIK 7P4
Canada

Visiting address: Pavillon De Koninck, bureau 2267, Faculté des Lettres, Université Laval
Head: Mr. Guy Rondeau, professeur titulaire
Staff categories:
3 terminologist(s) (special field); 1 documentalist(s)/librarian(s); 2 terminology teacher(s)
Founded: 1978
Documentation:
bibliographic data
factographic data .
Subject(s): all fields
UDC 0/9
Language(s): Fr; En; De; Ru
Vocabularies:
Subject(s): all fields
UDC 0/9
Language(s): Fr; En; De; Ru
Theory: Fr; En
Training: undergraduate; graduate; postgraduate
Access to training: Available to registered students only
Curriculum: fundamentals of terminology; terminological principles; lexicographical methods
Publications: Rondeau, G.; Felber, H. (ed.) *Textes choisis de terminologie. Vol. 1, Fondements théoriques de la terminologie*. Sous la direction de V. I. Siforov. Québec: GIRSTERM/Université Laval, 1981, 334 p., 140x190 mm, ISBN 2-90242-00-8. Obtainable through: GIRSTERM Faculté des lettres Université Laval Québec, G1K 7P4, Canada
RONDEAU, G.; FELBER, H. *Bibliographie internationale de la terminologie. Exposé de la méthode.* 2e éd. Québec: GIRSTERM/Université Laval, 1981, 68 p., 215x280 mm, ISBN 2-920242-02-4.
RONDEAU, G.; FELBER, H. et al. *Bibliographie de la terminologie, première tranche: 1300 titres classés par thèmes*;
RONDEAU, G. (ed.) Actes du *Colloque de février 1976* (version française);
RONDEAU, G. (ed.) *Actes du Colloque de février 1976* (version anglaise);
ROBERTS; SIMARD. *Langues de spécialités* no. 1; DUQUET-PICARD, D. *L'activité terminologique au Québec et à Ottawa*, 1980
Services: Consultation and bibliographies
No service charges
Operation:
computer-assisted
Data processing:
Hardware: 3031; 192 Kb; 1 Mb; 1 tape unit; printer with French characters; 2741 IBM
Software: operation mode: real time; batch; programming language(s): PLI and OS; data base management system/retrieval documentation system: APL and OS/Card file TSIO
Access: Possible

(Project(s))
Title: Analyse des caractères intrinsèques des terminologies utilisées dans l'implantation du français dans une entreprise [Analysis of the intrinsic characters of terminologies in the process of implementing French as a working language]
Project type: institutional; in progress
Principal investigator: Prof. Dr. Guy Rondeau
Collaborators: 3
Method of work: (1) Laboratory work: preparation of a grid for analysis; preparation of a questionnaire (2) Field work: analysis of the validity of the grid through users via presentation of the questionnaire

Title: Bibliographie internationale de la terminologie (BT1)/International bibliography of terminological literature (BT1)
Project type: commissioned project: (Bureau des traductions - Secrétariat d'Etat du Canada); in progress
Principal investigator: Prof. Dr. Guy Rondeau, Prof. H. Felber (Infoterm)
Collaborators: 3
Scope and/or subject matter: Terminology science; terminology work; terminography
Method of work: Collection and analysis of all bibliographical references of works pertaining to terminology
Source material: Reports; studies; papers; articles
Publications: Rondeau, G.: Felber, H. *Bibliographie internationale de la terminologie. Exposé de la méthode.* Québec: GIRSTERM/Université Laval, 1981, 68p., 215x280 mm, ISBN 2-920242-02-4. Publication of the Bibliography foreseen for 1983.; Research report on the international bibliography of terminology

Title: Constitution d'un dépôt des thèses et mémoires en terminologie au Canada [Compilation of theses and dissertations in terminology in Canada]
Project type: commissioned project:; in progress
Principal investigator: Prof. Dr. Guy Rondeau, Ms. Rosy Alber-DeWolf
Collaborators: 3
Scope and/or subject matter: This compilation includes related fields such as translation, lexicology and lexicography

Title: Critères de distinction du terme et du non-terme [Criteria of distinction between term and non-term]
Project type: thesis, dissertation; in planning
Principal investigator: Prof. Dr. Guy Rondeau
Collaborators: 1
Scope and/or subject matter: Theory of terminology

Title: Etude comparée de la création lexicale terminologique en français, en anglais, en allemand et en russe dans le domaine de la chimie [Comparative Study of lexical terminological creativity in French English, German and Russian in the field of chemistry]
Project type: thesis, dissertation (doctoral dissertation); in progress
Principal investigator: Prof. Dr. Guy Rondeau
Collaborators: 1
Scope and/or subject matter: Analysis of term-formation in the four languages in the field of spectroscopy as it applies to chemistry
Method of work: Statistical survey (by computer) of term-formation followed by a comparative analysis of derivability (prefixal and suffixal productivity) and a comparative analysis of creativity (formation of complex terminological units) in French, English, German, and Russian.

Title: Etude comparée de la définition lexicographique, encyclopédique et terminologique [Comparative study of the lexicographical, encyclopaedic and terminological Definition]
Project type: thesis, dissertation; in planning
Principal investigator: Prof. Dr. Guy Rondeau
Collaborators: 1

Title: Étude comparée du lexique forestier anglais-français-espagnol [Comparative study of forest terminology in En, Fr and Es]
Project type: thesis, dissertation; in progress
Principal investigator: Prof. Dr. Guy Rondeau
Collaborators: 1
Scope and/or subject matter: The main objective is to provide equivalent terms to concepts of forest terminology in En, Fr and Es
Method of work: The method of collecting and processing of terms is in conformity with a methodology considered appropriate by international organizations. It is systematic and represented by a tree covering the concepts of this field
Source material: Wüster, E. Dictionnaire de la machine-

outil (1968); ISO R/1149 Présentation des vocabulaires multilingues (1969); Auger, P et al. Méthodologie de la recherche en terminologie (1978)

Title: Etude liguistique du système de notions de la terminologie comme discipline [Linguistic study of the system of concepts in terminology as a discipline]
Project type: thesis, dissertation; in progress
Principal investigator: Prof. Dr. Guy Rondeau
Collaborators: 1
Scope and/or subject matter: General theory of terminology
Method of work: (A) Study and compare the different classification systems in existence; (B) establish a terminology tree on the basis of data collected
Source material: Boutin-Quesnel, R et al. Vocabulaire systématique de la terminologie. Ed. provisoire. Québec, éditeur officiel, 1978, 87 p.; Rondeau, G. Introduction à la terminologie. Edition provisoire. Québec: GIRSTERM Université Laval, 1980, 432 p.; Rondeau, G. Rapport de recherche sur la bibliographie internationale de la terminologie. Edition provisoire. Québec: GIRSTERM; Wüster, E. Introduction à la théorie générale de la terminologie et à la lexicographie terminologique. (Traduit de Einführung in die allgemeine Terminologielehre und terminologische Lexikographie). Québec: GIRSTERM, Université Laval (traduction française: Bureau des traductions du Canada), 1979, 87 p.

Title: Etude terminologique des moyens audio-visuels en milieu d'enseignement: aspect néonymique [Terminological study of audiovisual means in teaching: the neonymic aspect]
Project type: Office de la langue française du Québec; in planning
Principal investigator: Mr. J. -C. Boulanger
Collaborators: 3
Scope and/or subject matter: Audiovisual means in terminology; teaching; neonymy
Method of work: Methods used are those employed by the Office de la langue françcaise

Title: La synonymie dans les langues de spécialité (provisoire) [Synonymy in LSP (provisional)]
Project type: thesis, dissertation; in progress
Principal investigator: Prof. Dr. Guy Rondeau
Collaborators: 1
Scope and/or subject matter: Terminology (theoretical study)

Title: L'activité terminologique au Québec et à Ottawa [Terminological activities in the province of Quebec and in Ottawa]
Project type: thesis, dissertation
Principal investigator: Prof. Dr. Guy Rondeau
Collaborators: 1
Scope and/or subject matter: Terminology

Title: Le découpage du terme. Contribution à une problématique de la systématisation et de la méthode en terminologie [Delimitation of the term. Contribution to the problem of systematization and method in terminology]
Project type: thesis, dissertation; in progress
Principal investigator: Prof. Dr. Guy Rondeau, directeur de la thèse
Scope and/or subject matter: Theory and method of terminology (epistemology). Reflections on theoretical implications of concept analysis in relation to the term and examination of the conceptual process of equating concept and naming of a concept. This research allows for a theoretical and methodological framework concerning the analysis of problems in delimiting and isolating the term in terminology work. The second part covers the delimitation of simple and complex terminology units
Method of work: Work methods taken from linguistics and terminology with an approach borrowed from epistemology
Source material: Rondeau, G. (ed.) Table ronde sur les problèmes du découpage du terme (26 août 1978), Ve congrès de l'Association internationale de linguistique appliquée, Montréal, Commission de terminologie de L'AILA, Québec: OLF, Editeur officiel du Québec.
Publications: POTVIN D. *Analyse de la famille lexicale de terminologie, Travaux de terminologie no. 2.* Québec: éditions GIRSTERM/Université Laval, 1981

Title: L'emprunt linguistigue et l'intégration de certaines institutions anglo-américaines en droit québecois [Linguistic borrowing and the integration of certain Anglo-American institutions into the Quebec legal system]
Project type: thesis, dissertation (doctoral dissertation); in progress
Principal investigator: Prof. Dr. Guy Rondeau; Mr. Jean Goulet, Prof. of law (advisor)
Collaborators: 1
Scope and/or subject matter: This dissertation delves into linguistic borrowing in the field of economic and legal institutions that have been integrated into the framework of Quebec substantive law. It involves a description of those features peculiar to the vocabulary of such institutions that Quebeckers have translated into their language and culture thereby transformed them with respect to their Anglo-American cultural model. The study also comprises an examination of linguistic phenomena underlying the borrowing process as well as a description of the linguistic gap separating the Quebec and Anglo-American version of such institutions (i.e.corporate law, fiduciary relationships, sureties etc.)
Method of work: Research shall be carried out via on-line computerized interrogation of the Quebec Revised Statutes and the study shall be focused primarily upon the French version of the law which covers some 3 000 pages. The investigation involves seven steps: 1. Selection of terms taken from a comprehensive listing of lexical occurrences contained in the Quebec Statute Bank. Extraction of basic concepts from previous research carried out at the Office de la langue française. 2. Preparation of a conceptual tree structure for Quebec substantive law in those areas of interest to this dissertation. 3. Printing of KWIC lists with a view to textual analysis of concepts. 4. Analysis of KWIC lists in order to extract pertinent contexts having specific semantic features. Matching of contexts with the tree structure and changing it wherever necessary. 5. Systematic study of concepts within the framework of contextual limits. 6. Re-assessment of the basis of the theory of linguistic borrowing in the light of specialized languages, taking into account presently accepted socio-linguistic factors. Comparison with traditional theories on borrowing which are based more upon the whole of language than upon specialzed languages. 7. Writing up a synthesis of the preceding six steps.
Source material: 1. Le Recueil des lois du Québec, c. 1 - 394, Québec, Editeur officiel du Québec. 2. Schwab, W. "Les entreprises commerciales", in Meta, journal des traducteurs, mars 1979, numéro spécial sur la traduction juridique. 3. Schwab, W. "Pour qu'une traduction juridique soit légale." in La traduction une profession, dans les Actes du VIIIe Congrès mondial de la Fédération internationale des traducteurs, Montréal, 1977, p. 117–129. 4. Schwab, W. et Auger, P. "Le rôle de la néologie dans l'application des lois à caractère linguistique" dans Les implications de l'intervention juridique de l'Etat dans le domaine de la langue, Québec, Office de la langue française, 1977. 5. Schwab, W. "La réalisation du Thésaurus-S et du Thésaurus-G". In Thémis, la revue juridique de l'Université de Montréal, Montréal, 1971, octobre, p. 69–81.
Publications: 1. *La fiducie en droit québécois* (The Trust in the Quebec Legal System). A 25- page article that is scheduled for publication in the spring of 1982.
2. *Petit lexique du crédit-bail ou du Third-Party Leasing* [A

Mini Lexicon on Third-Party Leasing]. A ten page article now being drafted.
3. *Les anglicismes en droit positif québécois* [Anglicisms in Quebec Substantive Law.] A hundred page monograph that is planned for the summer of 1982.

Title: L'enseignement des langues de spécialité [The teaching of LSP]
Project type: thesis, dissertation; in progress
Principal investigator: Prof. Dr. Guy Rondeau
Collaborators: 11
Scope and/or subject matter: Linguistics

Title: Mémoire en terminologie de l'économétrie [Essay in the terminology of econometrics]
Project type: thesis, dissertation
Principal investigator: Prof. Dr. Guy Rondeau
Scope and/or subject matter: English-French terminology used in the field of model specification
Method of work: Method based on text analysis, delimiting of the field; structuring of concepts; research into equivalent terms on the basis of text analysis and the creation of neologisms, if necessary (aspects of standardization are not considered)
Source material: General and specialized vocabularies; specialized journals; specialized handbooks of econometrics

Title: Problèmes de la normalisation terminologique dans des pays différents mais partageant une langue commune (aspect socio-linguistique et communicationnel) [Problems of terminology standardization in different countries sharing a common language (socio-linguistic and communication aspects)]
Project type: thesis, dissertation; in planning
Principal investigator: Prof. Dr. Guy Rondeau
Collaborators: 1
Scope and/or subject matter: Electrical engineering
Method of work: Not yet defined
Source material: Not yet defined

Title: Textes choisis de terminologie (1er cycle) [Selected readings in terminology (B.A.level)]
Project type: institutional; in planning
Scope and/or subject matter: Terminology teaching

● ***491***(*Organization*)
IBM Canada Ltée
IBM; IBM; IBM
5, Place Ville-Marie Phone: (514) 874–8010
Montréal, Québec, H3B 2G3 Telex: 0525886
Canada
Visiting address: 5, Place Ville-Marie, Room 630, Montreal, Quebec, Canada
Staff: 40 (permanent employees)
Head: Mr. P. Glasheen, Manager of Communications
Expert(s): Ms. Nathalie Cartier, Manager, Terminology and Language Programs
Staff categories:
3 terminologist(s) (theory); 1 documentalist(s)/librarian(s); 13 translator(s); 23 other professional(s) (16 administrative personnel; 3 proof readers; 4 managers)
Founded: 1973
Documentation:
bibliographic data
factographic data
Subject(s): data processing; management; applied sciences; French language; English language
UDC 681. 3; UDC 351; UDC 80
Language(s): Fr; En
Vocabularies:
Subject(s): data processing; word processing; management; marketing; electronics; mechanical engineering; titles and functions
UDC 351; UDC 6
Language(s): En; Fr
Publications: *Posters on Magnetic Media*; *Customer Engineering Glossary* (poster of 500 terms); *Standard Product Line Glossary* (data processing); *Magnetic Media Glossary* (disk, card, tape, diskette, mass storage); *Termino*, a terminology bulletin, published thrice a year; *Selectric Typewriters Lexicon*; *Glossary of Terms Related to IBM Products and General Systems*; *Vocabulary of Magnetic Media*
Services: reference; translation; other: Production services
Service charges
User aids: Newsletter; Posters; Glossaries
Operation:
manual (in the process of being computerized)
Data base description:
Name: IBM Canada Terminology Database
Size: 11 000 entries
Language(s): En; Fr
Classification: Keyword indexing
Terminological data:
definitions; context; equivalents; other data elements (identification of language, source code and source for concept and equivalent, cross references, synonyms with source identification, name of author, date)
Data processing:
Hardware: IBM; 370; display stations, printers
Software:
operating system: VM (STAIRS); operation mode: real time
Access:
via telephone
Data available:
on exchange basis; free of charge

(Project(s))
Title: Vocabulaire des bases de données [Vocabulary of Data Bases]
Project type: institutional; in progress
Principal investigator: Mr. Richard Kromp, terminologist
Collaborators: 2
Scope and/or subject matter: Data processing
Method of work: (1) Terminological research; (2) Establishment of equivalences; (3) Production of glossary
Source material: Specialized documentation (books and periodicals)
Publications: The publication *Vocabulaire des bases de données* will be available upon request

● ***492***(*Commission/Committee*)
Comité interentreprises de traitement de textes (IBM, AES, OLF, Northern Telecom)
[IBM Canada Ltée]
CITT
Committee members: 5
Established: 1978
Vocabularies:
Subject(s): data processing; word processing
UDC 681. 3; UDC 80
Language(s): En; Fr
Vocabulary standards:
Subject(s): data processing; word processing
UDC 681. 3; UDC 80
Language(s): En; Fr

● ***493***(*Commission/Committee*)
CSA Committee on vocabulary
[IBM Canada Ltée]
Committee members: 15
Vocabularies:
Subject(s): data processing; word processing
UDC 681. 3; UDC 80
Language(s): En; Fr
Vocabulary standards:
Subject(s): data processing; word processing
UDC 681. 3; UDC 80
Language(s): En; Fr

● ***494*** *(Organization)*
Legal, Economic and Administrative Terminology Institute for Comparative Terminology
LEAT
40 Boteler Str., Suite 701
Ottawa, Ontario, K1N 9C8
Canada
Phone: (613) 238–4443
Telex: 053–4919 LEAT-OTT
Head: Dr. Alexander Lane, President
Expert(s): Dr. Alexander Lane
Founded: 1975
Documentation: bibliographic data
Subject(s): law; administration; economics
UDC 33; UDC 34; UDC 38
Language(s): All languages
Services: translation

● ***495*** *(Organization)*
Nekrassof, Dr. Vladimir, M. D. (private individual)
1218 Meadowlands Drive East, Suite 608
Ottawa, Ontario, K2E 6K1
Canada
Expert(s): Dr. Nekrassof Vladimir, M. D.

(Project(s))
Title: Dictionary of Biochemistry, English - German, German - English
Project type: Private undertaking; in progress
Principal investigator: V. N. Nekrassoff, M. D.
Collaborators: 3
Scope and/or subject matter: Biochemistry (incl. medical biochemistry) and methods used in biochemistry
Method of work: Literature search
Source material: Current world literature
Publications: Various glossaries in *Lebende Sprachen* and *Meta*

Title: Dictionary of Entomology. En - De, De - En
Project type: Private undertaking; in progress
Principal investigator: V. N. Nekrassof, M. D.
Collaborators: 3
Scope and/or subject matter: Entomology: Insect anatomy, physiology, ecology
Method of work: Literature search
Source material: Current world literature

Title: Dictionary of Ethology. German - English, English - German
Project type: Private undertaking; in progress
Principal investigator: V. N. Nekrassoff, M. D.
Collaborators: 3
Scope and/or subject matter: Terminology of animal behaviour
Method of work: Literature search
Source material: World literature

Title: Dictionary of Ornithology. En, Fr, De
Project type: Private undertaking; in progress
Principal investigator: V. N. Nekrassoff, M. D.
Collaborators: 3
Scope and/or subject matter: Birds: Their anatomy, physiology, behaviour, ecology
Method of work: Literature search
Source material: Current world literature

Title: Dictionary of Palearctic Birds. En, Fr, De
Project type: Private undertaking; in progress
Principal investigator: V. N. Nekrassoff, M. D.
Collaborators: 3
Scope and/or subject matter: Names of birds (species, genera, families, orders)
Method of work: Literature search
Source material: World literature

Title: Dictionary of the Vertebrates of North America. En, Fr, De
Project type: Private undertaking; in progress
Principal investigator: V. N. Nekrassoff, M. D.
Collaborators: 3
Scope and/or subject matter: Names of all mammals, birds, fish, reptiles and amphibians (species, families, orders)
Method of work: Literature search
Source material: Current literature

● ***496*** *(Organization)*
Office de la langue française - Direction de la terminologie
OLF
Parent organization: Conseil de la langue française; Commission de surveillance de la langue française; other organizations created by the "Charte de la langue française"
700, boul. St-Cyrille est,
2e étage, bureau 203
Québec, Québec, G1R 5G7
Canada
Phone: 643–2134
Telex: AFINTER QBC 051–3523 OLF
Visiting address: 700, boul. St-Cyrille est, 2e étage, bureau 203
Staff: 90
Head: Mr. Pierre Auger, Directeur de la terminologie
Expert(s): Mr. Louis-Jean Rousseau, chef du Service des travaux terminologiques; Ms. Tina Célestin, chef du Service des consultations; Mr. Jean-Claude Gaumond, chef du Service des linguistes-conseils
Staff categories:
9 terminologist(s) (special field); 28 terminologist(s) (theory); 30 linguist(s); 2 terminology teacher(s); 1 other professional(s)
Founded: 1961
Documentation:
bibliographic data
factographic data
Subject(s): all fields
UDC 0/9
Language(s): Fr; En
Vocabularies:
Subject(s): almost all fields (technology; industrial security; packaging; machine-tools; machining; tools; assembling; fluid mechanics; management; law; food; neology)
UDC 0/9
Language(s): Fr; En
Vocabulary standards:
Subject(s): all fields
UDC 0/9
Language(s): Fr
Theory: Fr
Training: undergraduate; graduate
Curriculum: fundamentals of terminology; terminological principles; lexicographical methods
Publications: Vocabularies concerning a large number of fields, namely: food, sports, insurance, cars, administration, technology, industry; Collection *Etudes, recherches et documentation*; Collection *Langues et sociétés*;*Bulletins de diffusion du français*; *Cahiers de néologie*; Rapports des colloques et réunions de terminologie; and other publications
Services: reference (via Banque de terminologie du Québec); loan (via Banque de terminologie du Québec); reproduction (Directorate of communications, OLF); referral (via Banque de terminologie du Québec); other: Linguistic and terminological counselling; francization
Service charges: On-line
No service charges: Off-line
User aids: "Terminogramme". Bulletin de la Direction de la terminologie; "En Marche" Bulletin de francisation de l'Office de la langue française, bi-monthly
Operation:
manual (terminology service)
computer-assisted (terminological information service)

Data base description: Use of equipment of the Banque de terminologie du Québec
Terminological data:
definitions; explanations; classified layout; context; standardized terms (depending on field); equivalents; other data elements
Data processing:
Software: Use of equipment of the Banque de terminologie du Québec
Access:
via telephone/via telex

(Project(s))
Responsible: Réseau de néologie scientifique et technique
Title: Bibliographie linguistique de la néologie [Linguistic bibliography in the field of neology]
Principal investigator: M. Jean-Claude Boulanger
Collaborators: 1
Scope and/or subject matter: Bibliography on linguistic studies in the field of neology (1960–1980); analytical bibliography of lexicographical works in the field of neology (1960–1980)
Method of work: Inventory and analysis of documents
Source material: Linguistic and terminological documentation; international linguistic bibliography; organisms and universities
Publications: Boulanger, J. -C. *Bibliographie linguistique de la néologie*: 1960–1980, I. *Etudes linguistiques*, coll. *Etudes, recherches et documentation*, Québec, Office de la langue française, Editeur officiel du Québec, November 1980, 300 p.;
Boulanger, J. -C. *Bibliographie linguistique de la néologie: 1960–1980, II. Lexicographie*, 1981 (in preparation); Bibliographical card-index

Title: Chroniques terminologiques dans la revue "Parlure" [Terminology section in the magazine "Parlure"]
Project type: institutional; in progress
Principal investigator: Mrs. Tina Célestin
Collaborators: 2
Scope and/or subject matter: Traditional crafts (ex.: blacksmithing, pipe making, weaving etc.)
Method of work: Standard terminological research procedure
Source material: General and specialized reference works as requested by the study undertaken
Publications: Already published: violin, bookbinding, blacksmithing, weaving, ceramics, pipe making; *Parlure*. Québec. Les Editions Parlure inc. Vol. 1, No. 1, 1976; to come: basketry and others yet undetermined

Title: Diffusion de fiches linguistiques et terminologiques dans l'Administration et dans le public [Issuance of linguistic and terminological documents to government agencies and to the public in general]
Principal investigator: Mrs. Tina Célestin
Collaborators: Undetermined
Scope and/or subject matter: Answers to most common linguistic and terminological problems
Method of work: Discussions on theoretical and practical aspects of term research: definition, goal, methodology, applications, etc.
Source material: Linguistic and terminology reference works
Publications: Format and publication date are yet to be determined

Title: Etude sur la recherche terminologique ponctuelle [Term research study]
Project type: institutional; in progress
Principal investigator: Mrs. Tina Célestin
Collaborators: Four-member committee
Scope and/or subject matter: Term research: definition and methodology
Method of work: Inventory of the most pressing linguistic and terminological needs issuing from government agencies and the public in general in order to provide answers
Source material: Any general and specialized reference work, as required
Publications: Publication format and date are yet undetermined

Title: Informatisation de la fiche de consultation [Computerization of the term research form]
Principal investigator: Mrs. Tina Célestin
Collaborators: Undetermined
Scope and/or subject matter: Compiling of statistical data
Method of work: Computerization of data collected on the term research forms

Responsible: Service des travaux terminologiques
Title: Méthodologie de la recherche terminologique [Methodology of terminological work]
Project type: institutional; in progress
Principal investigator: Mr. Louis-Jean Rousseau, chef du Service des travaux terminologiques
Collaborators: 5
Scope and/or subject matter: Up-dating of the document entitled "Méthodologie de la recherche terminologique" published in 1978 by the OLF
Method of work: A theoretical view on methodological work on the basis of practical experience (field work, reference comittees, etc.)
Source material: All specialized documentation available on the subject
Publications: Auger, P.; Rousseau, L. -J. *Méthodologie de la recherche terminologique*. Collection *Etudes, recherches et documentation*, Québec, Office de la langue française, 1978.

Responsible: Réseau de néologie scientifique et technique
Title: Methodology of research in the field of neology
Principal investigator: Mr. Jean-Claude Boulanger
Collaborators: 4
Scope and/or subject matter: Prepare a working methodology in the field of neology
Method of work: Based on the practical experience and on the theoretical works
Source material: Works of the "Réseau de néologie scientifique et technique"; terminological works
Publications: Auger, P.; Rousseau, L. -J. et al. *Méthodologie de la recherche terminologique*, coll. *Etudes, recherches et documentation*. Québec: Office de la langue française, Editeur officiel du Québec, May 1978, 80 p.;
Boulanger, J. -C. *Néologie et terminologie*, Néologie en marche, série b: Langues de spécialité, no. 4, Québec, Office de la langue française, Editeur officiel du Québec, September 1978, p. 5–127;
Boulanger, J. C. *Problématique d'une méthodologie d'identification des néologismes en terminologie*, Néologie et lexicologie. Hommage à Louis Guilbert, coll. *Langue et Langage*, Paris, Librairie Larousse, 1979. p. 36–46.;
Boulanger, J. -C. Bibliographie linguistique de la néologie: 1960–1980, I. Etudes linguistiques, coll. *Etudes, recherches et documentation*, Québec, Office de la langue française, Editeur officiel du Québec, November 1981, 300 p.;
Boulanger, J. -C. (coordinator), Néologie en marche (série b: langues de spécialité), Québec, Office de la langue française, Editeur officiel du Québec, cahiers 1 (septembre 1976) à 26 (mai); Néologie en marche (as of cahier 16)

Title: Programme d'échange de terminologues [Terminologists exchange programme]
Project type: institutional; in progress
Principal investigator: Mrs. Tina Célestin
Collaborators: Any service, terminologists may be offered to participate

Scope and/or subject matter: Exchanges may be arranged with organizations in the province of Québec, other provinces of Canada and other countries
Method of work: Schedule established according to visiting terminologists' wishes and expectations

Title: Recherches ponctuelles [Term research]
Project type: institutional; in progress
Principal investigator: Mrs. Tina Célestin
Collaborators: 21
Scope and/or subject matter: Answering approximately 60 000 requests a year submitted to the Service by phone, letters and on-location consultations. These requests deal with problems spanning from general linguistic matters to specific terminological questions in all fields of human activity
Method of work: Regular term research procedure
Source material: Computerized and conventional reference works (files, dictionaries, handbooks, standards, etc.); consulting experts in most fields of human activity

Title: Révision de textes [Revision of manuscripts]
Principal investigator: Ms. Hélène Martin
Scope and/or subject matter: Revision of manuscripts and any other written material originating mainly from l'Office de la langue française and occasionally from public or private organizations

Title: Séances d'animation dans les ministères [Information sessions for government employees]
Project type: institutional; in progress
Principal investigator: Ms. Diane Lambert-Tésolin
Collaborators: 1
Scope and/or subject matter: Information on the use of reference works
Method of work: Information sessions
Source material: Most common and available reference works

Responsible: Direction de la terminologie
Title: Terminogramme
Project type: institutional; in progress
Principal investigator: Mr. Bengt Lindfelt
Collaborators: Variable
Scope and/or subject matter: Newsletter on terminology: papers, scientific information; documentation
Method of work: Editing
Publications: *Terminogramme* no. 0 (October 1979) to 7/8 (March/June 1981)

● ***497*** *(Commission/Committee)*
Commission de terminologie de l'éducation
[Education Terminology Commission]
Parent organization: Office de la langue française (OLF)
316, Tour de la bourse Phone: (514) 873 5376
Montréal, Québec
H4Z 1GB
Visiting address: 800, place Victoria, 15e étage, Montréal
Contact: M. Pierre Auger, Directeur de la terminologie, Office de la langue française, 700 boul. St-Cyrille est, Québec G1R 5G7, Canada
Committee members: 15
Chairperson: Prof. Jean Denis Gendron
Established: 1980
Vocabularies:
Subject(s): education (public management)
UDC 37. 014
Language(s): En; Fr
Vocabulary standards: OLF has legal standardization authority. The Commission examines all standardization requests and submits advice to OLF
Subject(s): education (public management)
UDC 37. 014.55
Language(s): Fr
Publications: Commission's advices to OLF lead to the publication of standards in the Gazette officielle du Québec. These standards are to be collected and reprinted annually by the Editeur officiel du Québec (State Printer). They also appear in Terminogramme, the Terminology Branch (OLF) newsletter; Terminogramme newsletter is available free of charge through Mr. Pierre Auger

● ***498*** *(Commission/Committee)*
Commission de terminologie de l'Office de la langue française
C. T. O. L. F.
Parent organization: Office de la langue française (OLF)
700, boul. St-Cyrille est Phone: 418–643–2134
Québec, Québec, G1R 5G7
Canada
Contact: Rosita Harvey, Secrétariat de la Commission de terminologie, 700, boul. St-Cyrille est, Québec, Canada G1R 5G7
Committee members: 6
Chairperson: Mr. Pierre Auger, Directeur de la terminologie
Established: 1978
Vocabularies: The Commission prepares notices (recommendation or standardization) principally of terminologies elaborated by the Directorate of terminology
Subject(s): all fields
UDC 0/9
Language(s): Fr; En
Vocabulary standards: (The commission also prepares OLF policies on different projects of general interest such as linguistic borrowing, the status of regionalisms, criteria for standardization, etc.)
Subject(s): all fields
UDC 0/9
Language(s): Fr
Theory: Fr
Publications: Enoncé de politique relative à l'emprunt de formes linguistiques étrangères [Policy statement concerning the borrowing of foreign linguistic forms]; Répertoire des avis linguistiques et terminologiques de l'Office de la langue française 1979–1980 [Collection of linguistic and terminological notices issued by the Office de la langue française 1979–1980]; without charge

● ***499*** *(Commission/Committee)*
Commission de terminologie des Affaires sociales
[Social Affairs Terminology Commission]
Parent organization: Office de la langue française (OLF)
316, Tour de la bourse Phone: (514) 873–5376
Montréal, Québec, H4Z 1G8
Canada
Visiting address: 800, place Victoria, 15e étage, Montréal
Contact: M. Pierre Auger, Directeur de la terminologie, Office de la langue française, 700, boul. St-Cyrille est, Québec G1R 5G7, Canada
Committee members: 10
Established: 1978
Vocabularies:
Subject(s): public; health (management); social security; welfare; social services
UDC 362. 1; UDC 368. 4; UDC 361; UDC 351. 84
Language(s): En; Fr
Vocabulary standards: OLF has legal standardization authority. The Commission examines all standardization requests and submits advice to OLF
Subject(s): public health (management); social security; welfare; social services
UDC 362. 1; UDC 368. 4; UDC 351. 84;
Language(s): Fr
Publications: Commission's advices to OLF lead to the publication of standards in the Gazette officielle du Québec. These standards are to be collected and reprinted annually by the Editeur officiel du Québec (State Printer). They also appear in Terminogramme, the Terminology Branch (OLF) newsletter; Terminogramme newsletter is available free of charge through Mr. Pierre Auger

● ***500*** *(Commission/Committee)*
Commission de terminologie des Transports
[Transport Terminology Commission]
Parent organization: Office de la langue française (OLF)
316, Tour de la bourse Phone: (514) 873–5376
Montréal, Québec, H4Z 1G8
Canada
Visiting address: 800, place Victoria, 15e étage, Montréal
Contact: M. Pierre Auger, Directeur de la terminologie, Office de la langue française, 700, boul. St-Cyrille est, Québec G1R 5G7, Canada
Committee members: 6
Established: 1978
Vocabularies:
Subject(s): road design; road construction; public works equipment
UDC 625. 7. 05; UDC 351. 712. 4
Language(s): En; Fr
Vocabulary standards: OLF has legal standardization authority. The Commission examines all standardization requests in the field and submits advice to OLF
Subject(s): road design; road construction; public works equipment
UDC 625. 7. 05; UDC 351. 712. 4
Language(s): Fr
Publications: Commission's advices to OLF lead to the publication of standards in the Gazette officielle du Québec. These standards are to be collected and reprinted annually by the Editeur officiel du Québec (State Printer). They also appear in Terminogramme, the Terminology Branch (OLF) newsletter; Terminogramme newsletter is available free of charge through Mr. Pierre Auger

● ***501*** *(Commission/Committee)*
Commission de terminologie géographique
[Commission on Geography Terminology]
Parent organization: Office de la langue française (OLF)
316, Tour de la bourse Phone: (514) 873–5376
Montréal, Québec, H4Z 1G8
Canada
Visiting address: 800, place Victoria, 15e étage, Montréal
Contact: M. Pierre Auger, Directeur de la terminologie, Office de la langue française, 700, boul. St-Cyrille est, Québec G1R 5G7, Canada
Committee members: 8
Established: 1979
Vocabularies:
Subject(s): geography
UDC 91
Language(s): En; Fr
Vocabulary standards: OLF has legal standardization authority. The Commission examines all standardization requests and submits advice to OLF
Subject(s): geography; toponymy
UDC 91
Language(s): Fr
Publications: Commission's advices to the OLF lead to publication of standards in the Gazette officielle du Québec. These standards are to be collected and reprinted annually by the Editeur officiel du Québec (State Printer). They also appear in Terminogramme, the Terminology Branch (OLF) newsletter; Terminogramme newsletter is available free of charge through Mr. Pierre Auger

(Project(s))
Responsible: Commission de terminologie de l'Office de la langue française
Title: Critéres de recevabilité des dossiers soumis à la Commission [Admissibility criteria requested of documents submitted to the Commission]

Responsible: Commission de terminologie de l'Office de la langue française
Title: Enoncé de politique en matière de créativité lexicale [status of neologisms]

Responsible: Commission de terminologie de l'Office de la langue française
Title: Enoncé de politique sur les régionalismes (status of regionalisms)

● ***502*** *(Data Bank)*
Office de la langue française - Direction de la banque de terminologie du Québec
[French Language Board - Directorate of the terminology bank of Québec (not official)]
OLF - BTQ
Parent organization: Conseil de la langue française - Commission de surveillance de la langue française
700 boulevard Saint-Cyrille
est, 2°e étage Phone: (418) 643–1802
Québec, Québec, G1R 5G7 Telex: AFINTER QBC 051–3523 OLF
Canada
Staff: 71
Head: M. Jean-Marie Fortin, Directeur de la Banque de terminologie du Québec
Expert(s): M. Buddhi Klok, Service de la documentation; Mme Francine Thomas, Groupe d'étude et de coordination
Staff categories:
31 terminologist(s) (theory); 5 documentalist(s)/librarian(s); 23 information specialist(s); 12 other professional(s)
Founded: 1961 (OLF), 1973 (BTQ)
Purpose of data bank/users: Diffuse terminology in respect
377, 378, 388, 393, 425, 427, 422, 454, 467, 472, 514
translators, teachers, firms
Operation:
computerized
Data base description:
Name: Banque de terminologie du Québec
Size: + /- 830 000 documents (Terminologie); + /- 11 000 doc. (Sources); + /- 29 000 doc. (Inventaire)
Language(s): Fr; En
Classification: Thesaurus; UDC; data base; dictionary
Subject(s): all fields
UDC 0/9
Language(s): Fr; En
Terminological data:
definitions; explanations; classified layout; context; equivalents; other data elements
Products: Word lists; glossaries; magnetic tapes, disks (on exchange basis only)
Service charges (on-line)
No service charges (off-line)
Access:
via telephone / via telex
on-line; off-line
Data available:
on exchange basis
Data processing:
Hardware/Capacity: Amdhal; HP (Hewlett Packard) V6; 3 000; 8 mb (X3); 768Kb; -; 240 000 000 caractères; 6 250 PBI (X); 1 600 BPI; 1 printer, 13 terminals; 18 terminals; 13 terminals
Software:
operating system: MVS 3. 8 JES 2; PME III; operation mode: BATCH, REAL TIME, DEMAND, MULTIPLE REOMOTE JOB ENTRY; BATCH, REAL TIME, MULTIPLE REMOTE JOB ENTRY; programming language(s): PL1, COBOL, ASSEMBLER, RPG, FORTRAN, SPL; data base management system/retrieval documentation system: CICS/STAIRS

● ***503*** *(Organization)*
School of Translators and Interpreters
[Ecole de traducteurs et d'interprètes]
ETI
Parent organization: University of Ottawa
Ottawa, K1N 6N5 Phone: (613) 231–2346
Canada
Staff: On terminology work: 3

Head: Dr. R. Roberts
Expert(s): Dr. J. Delisle
Staff categories:
2 terminologist(s) (special field); 1 terminologist(s) (theory); 1 other professional(s) (computer specialist)
Founded: 1970
Documentation: bibliographic data
Subject(s): law; interpreting
UDC 34; UDC 82. 03
Language(s): En; Fr
Vocabularies:
Subject(s): law; interpreting
UDC 34; UDC 82. 03
Language(s): En; Fr
Theory: Fr
Training: undergraduate; graduate
Curriculum: fundamentals of terminology; terminological principles; lexicographical methods. Undergraduate: 1 course on fundamentals of terminology; terminological principles; lexicographical methods; 1 course on computer aids for translation; covers terminology banks. Graduate: 1 seminar on theoretical foundations of terminology; 1 seminar on legal terminology
Publications: DELISLE, J.; ALBERT, L. *Bibliographic Guide for Translators, Writers and Terminologists (Cahiers de traductologie* no.1), Ottawa, Editions de l'Université d'Ottawa, 1979.

○ ***504*** *(Organization)*
Standards Council of Canada. International Standardization Branch
SCC
2000 Argentia Road,
Suite 2–401 Phone: 1 416 826 8110
Mississauga, Ontario,
L5N 1V8 Telex: 06–97719 scc msga
Canada Telegr.: intstancan toronto
Head: Mr. J. R. Woods, Executive Director
Vocabulary standards:
Subject(s): all fields
UDC 0/9
Language(s): En; Fr

● ***505*** *(Organization)*
Université de Montréal, Département de linguistique et philologie
C.P. 6128, Succursale "A" Phone: (514) 343–6220
Montréal, H 3C 3J7
Canada

(Project(s))
Responsible: Coopération France - Québec
Title: Mini-banque de terminologie spécialisée [Specialized Terminology Mini-bank]
Project type: commissioned project:; in progress
Principal investigator: Jean Baudot; André Clas; Maurice Gross
Collaborators: 7
Scope and/or subject matter: Use of mini-computer to interrogate highly specialized bilingual (En/Fr) terminology (present field: solar energy)
Method of work: Emphasis is on data organization to facilitate access to complex entries
Source material: English-French Solar Energy Vocabulary
Publications: Working papers

○ ***506*** *(Organization)*
Dirección general de normas
[General Office of Standards]
DGN
Calle Puente de Tecamachalco N.U 6, Lomas de Tecamachalco, Sección Fuentes, Naucalpan de Juárez
Mexico, 53950 Telex: sepafin 1775690 duiame
Mexico Telegr.: sepafin-36
Head: Dr. Román Serra, Director General
Vocabulary standards:
Subject(s): all fields
UDC 0/9
Language(s): Es

○ ***507*** *(Organization)*
Comité estatal de normalización
[State Committee of Standardization]
NC
Egido 602 entre Gloria y Apodaca
Zona postal 2 Phone: 53 7 6 7901
La Habana
Cuba
Head: Mr. R. Darias Rodés
Vocabulary standards:
Subject(s): all fields
UDC 0/9
Language(s): Es

○ ***508*** *(Organization)*
Jamaica Bureau of Standards
JBS
P. O. Box 113 Phone: 809 9026 31 40
Kingston 10
Jamaica Telegr.: stanbureau jamaica
Head: Dr. A. S. Henry, Director
Vocabulary standards:
Subject(s): all fields
UDC 0/9
Language(s): En

○ ***509*** *(Organization)*
Dirección general de normas y sistemas de calidad. Secretaría de industria y comercio
[General Directorate of Standards and Systems of Quality, Secretary of Industry and Commerce]
DIGENOR
Av. Mexico No. 30 Phone: 566 39 20
132 Santo Domingo, Dominican Republic
Head: Mr. Arif Abud Abreu, Director General
Vocabulary standards:
Subject(s): all fields
UDC 0/9
Language(s): Es

○ ***510*** *(Organization)*
Barbados National Standards Institution
BNSI
"Flodden" Culloden Road
St. Michael
Barbados
Vocabulary standards:
Subject(s): all fields
UDC 0/9
Language(s): En

● ***511*** *(Organization)*
Agnew Tech-Tran Inc.
ATT
21050 Erwin Street
P. O. Box 789 Phone: (213) 340–5147
Woodland Hills,
California, 91365 Telex: 181524
USA
Staff: 297
Head: Ms. Irene Agnew
Expert(s): Mr. Fred Klein
Staff categories:
1 documentalist(s)/librarian(s); 1 information specialist(s); 231 translator(s) (13 full time; 218 part time); 207 other professional(s) (typesetters, word processing specialists, narrators, interpreters, graphic artists)

Founded: 1964
Documentation:
bibliographic data
factographic data
Subject(s): computers; aviation; missiles; ammunition; oil (plus a variety of many others)
UDC 622; UDC 623; UDC 629. 7; UDC 681. 3
Language(s): Ar; Ch; De; En; Fr; Ru; Es
Vocabularies:
Subject(s): computers; aviation; missiles; ammunition; oil
UDC 622; UDC 623; UDC 629. 7; UDC 681. 3
Language(s): Ar; Ch; De; En; Fr; Ru; Es
Publications: Multilingual glossaries and word lists
Services: translation; other: (multilingual glossaries/word lists; complete simultaneous interpretation services; camera-ready brochures, etc.in most foreign languages; telecopier services)
Service charges
User aids: Newsletter
Operation:
computer-assisted (combining Mergenthaler Linotype's MVP input/editing terminal and V-I-P phototypesetter))
Data base description:
Name: ATT Word Bank
Language(s): Ar; De; En; Fr; Ru; Es
Classification: SIC
Terminological data:
equivalents
other data elements
Access:
via telex
Data available:
on exchange basis; with charge

(Project(s))
Title: Glossary of English and Russian Computer and Automated Control Systems Terminology
Project type: institutional
Principal investigator: Ms. Irene Agnew
Scope and/or subject matter: Computer science; electronic data processing
Method of work: The glossary contains over 3000 terms and phrases in En and Ru - compiled in the course of translating a large number of computer industry projects from Ru into En and vice versa. Also included is a list of abbreviations (in En and Ru)

● ***512*** *(Organization)*
American Medical Association
AMA
535 North Dearborn Street Phone: (312) 751–6776
Chicago, Illinois, 60610
USA
Staff: 8
Head: James H. Sammons, M. D., Executive Vice-President
Expert(s): Joseph B. Jerome, Ph. D.
Staff categories:
1 terminologist(s) (special field); 1 documentalist(s)/librarian(s)
Founded: 1847
Documentation:
bibliographic data
factographic data
Subject(s): medicine (and related fields)
UDC 61
Language(s): En
Publications: FINKEL, Asher J. (ed.) *Physicians' Current Procedural Terminology*. Fourth Edition; FINKEL, Asher J. (ed.) *Current Medical Information and Terminology*. Fifth Edition
Operation:
computerized
Terminological data:
definitions; explanations; standardized terms

(Project(s))
Title: Current Medical Information and Terminology
Project type: institutional; in progress
Principal investigator: Asher J. Finkel, M. D.
Collaborators: Numerous
Scope and/or subject matter: System adopted to name and describe diseases for reference in clinical recording and reporting
Method of work: Consultation; medical library research
Source material: Medical textbooks; medical journals; consultants representing medical specialities
Publications: *Current Medical Information and Terminology*

Title: Physicians' Current Procedural Terminology
Project type: institutional; in progress
Principal investigator: Asher J. Finkel, M. D.
Collaborators: Numerous
Scope and/or subject matter: Systematic listing and coding of procedures and services performed by physicians
Method of work: Medical specialty representatives evaluate additions and suggestions for change with judgement by Ad Hoc Editorial Consultant Panel
Source material: Medical Specialty Society representatives; medical publications
Publications: *Physicians' Current Procedural Terminology*

○ ***513*** *(Organization)*
American National Standards Institute
ANSI
1430 Broadway Phone: 1 212 345 3300
New York, N. Y. 10018 Telex: 42 42 96 ansi ui
USA Telegr.: standards, new york
Head: Mr. Donald L. Peyton, Executive Vice-President
Vocabulary standards:
Subject(s): all fields
UDC 0/9
Language(s): En

● ***514*** *(Commission/Committee)*
Committee on Terminology
COT (S 18)
Parent organization: American Society for Testing and Materials
1916 Race Street Phone: (215) 299–5505
Philadelphia, Pennsylvania, 19103
USA
Contact: Ms. Kathleen Riley
ASTM, 1916 St., Philadelphia
Pa. 19103, USA
Committee members: 1 staff member; 15 committee members; 137 liaison members
Chairperson: Ms. Nancy M. Trahey
Vice-chairperson: Mr. Wayne P. Ellis
Established: 1920
Vocabulary standards:
Subject(s): materials testing; technology
UDC 5/6; UDC 620. 1
Language(s): En
Publications: *Compilation of ASTM Standard Definitions*. Fourth Edition. Philadelphia: ASTM, 1979, 792 p., 150x225. This new volume which is a major update of the 1976 edition contains over 14, 000 definitions from ASTM standards. The book will be of great value to those requiring the use of definitions of technical terms, especially those not commonly found in dictionaries.

● ***515*** *(Organization)*
Educational Resources Information Center, ERIC Processing and Reference Facility
ERIC
Parent organization: National Institute of Education (NIE)
4833 Rugby Avenue, Suite 303 Phone: (301) 656–9723
Bethesda, Maryland, 20814
USA

Staff: 1
Head: Mr. W. T. Brandhorst
Expert(s): Mr. J. Houston, Lexicographer
Staff categories:
1 documentalist(s)/librarian(s)
Founded: 1966
Documentation:
bibliographic data
factographic data
Subject(s): education (and related fields)
UDC 37
Language(s): En
Vocabularies:
Subject(s): education (and related fields)
UDC 37
Language(s): En
Vocabulary standards:
Subject(s): education (and related fields)
UDC 37
Language(s): En
Theory: En
Publications: Educational Resources Information Center. *Thesaurus of ERIC Descriptors*, Completely revised, 1980. Phoenix, Arizona: Oryx Press, 419 p. ($19.75 USA; $20.95 other countries). Lists more than 8000 terms. Over 650 new indexable terms are included in the revised *Thesaurus*. Additionally, over 1, 000 terms listed in earlier editions have been deleted. Postings of most of the deleted terms have been transferred to other terms. These deleted terms are, therefore, carried as synonyms in the revised Thesaurus. Also, approximately 1, 400 new or modified Scope Notes to older terms have been added. Next edition available early 1982 for approx. $25.00.
Bibliographies and other products produced from the Data Base by the ERIC Clearinghouses are announced monthly in the abstract journal *Resources in Education* (RIE)
Services: other: Centralized lexicographic analysis and subject referral service provided for all ERIC indexers and searchers
Operation:
computerized
Data base description:
Name: ERIC
Size: 400,000 documents and journal articles, indexed with "Thesaurus of ERIC Descriptors" (5,000 main indexable terms and 3,500 synonyms) and "ERIC Identifier Authority List" (approximately 26,000 unstructured, non-Thesaurus concepts)
Language(s): En
Classification: "Thesaurus of ERIC Descriptors"; "Identifier Authority List"
Terminological data:
definitions; explanations; classified layout; context
Data processing:
Hardware: ITEL; AS/5–I; AS/6–II (IBM 370/168 equivalent)
Software:
operating system: ERIC Software (internal system used only by ERIC); retrieval documentation system: Access to ERIC Data Base is provided by external search systems
Access:
via telephone
on-line; off-line
Data available:
with charge (details available from ERIC)

● ***516*** *(Data Bank)*

Educational Resources Information Center, ERIC Processing and Reference Facility
ERIC
Parent organization: National Institute of Education (NIE), Washington, D. C
4833 Rugby Avenue, Suite 303 Phone: (301) 656–9723
Bethesda, Maryland, 20814
USA
Staff: 1
Head: Mr. W. T. Brandhorst
Expert(s): Mr. J. Houston, Lexicographer
Staff categories:
1 documentalist(s)/librarian(s)
Founded: 1966
Purpose of data bank/users: Information storage and retrieval in the field of education; Educational policymakers, administrators, instructional personnel, school supportive personnel, researchers, students, parents, community members, etc.
Operation:
computerized
Data base description:
Name: ERIC
Size: 400,000 documents and journal articles, indexed with 5, 000 Descriptors (Thesaurus terms) and 26, 000 Identifiers (unstructured, non-Thesaurus terms)
Language(s): En
Classification: "Thesaurus of ERIC Descriptors, Completely Revised, 1980". (Available from Oryx Press, 2214 N.Central Avenue at Encanto, Phoenix, Arizona 85004, $ 19.75 USA, $ 20.95 other countries). Next edition available early 1982 for approx. $ 25.00
Subject(s): education (and related fields)
UDC 37
Language(s): En
Terminological data:
definitions; explanations; classified layout; context
Products: For details, see the following documents, available free from the ERIC Processing and Reference Facility: "How to Use ERIC" ; "How to Start an ERIC Collection"; "Directory of ERIC Search Services"; "Directory of ERIC Microfiche Collections"
Access:
via telephone
on-line; off-line
Data available: Details in "Directory of ERIC Microfiche Collections"
with charge
Data processing:
Hardware/Capacity: ITEL AS/5–I; AS/6–II (IBM 370/168 equivalent)
Software:
operating system: ERIC Software (internal system used only by ERIC); retrieval documentation system: Access to ERIC Data Base is provided by external search systems

● ***517*** *(Commission/Committee)*

Office of Biochemical Nomenclature Biology Division, Oak Ridge National Laboratory (since 1977)
OBN
Parent organization: National Academy of Sciences - National Research Council (USA) (1965–1977)
Box Y Phone: (615) 574–0808
Oak Ridge, Tennessee, 37830
USA
Chairperson: Dr. Waldo E. Cohn, Director
Established: 1965
Vocabularies:
Subject(s): biochemistry; organic chemistry
UDC 577. 1; UDC 547
Language(s): En
Publications: Reprints of international rules and recommendations (IUPAC, IUB); summaries of above plus recommendations emanating from this office in areas not covered by above; with out charge

● ***518*** *(Organization)*

United States Adopted Names Council
USAN
Parent organization: American Medical Association; U. S. Pharmacopeial Convention, Inc.; American Pharmaceutical Association
535 North Dearborn Street Phone: (312) 751–6565
Chicago, Illinois, 60610
USA
Staff: 4

Head: John Andrako, Ph. D., Chairman
Expert(s): USAN (United States Adopted Names Council) members: John Andrako, Ph. D.; Stanley Koch, M. S.; Lloyd Miller, Ph. D.; Donn Smith, M. D.; Lauren Woods, M. D.
Staff categories:
7 terminologist(s) (theory and spezial field); 2 documentalist(s)/librarian(s); 1 information specialist(s)
Founded: 1964
Documentation: factographic data
Subject(s): drugs; pharmaceutical products (non-proprietary nomenclature for drugs and related pharmaceutical ingredients)
UDC 615
Language(s): En
Vocabularies:
Subject(s): drugs (nomenclature)
UDC 615
Language(s): En
Vocabulary standards:
Subject(s): drugs (nomenclature)
UDC 615
Language(s): En
Theory: En
Publications: *USAN and the USP Dictionary of Drug Names* (1980/81); JAMA *New Names* - monthly column in the Journal of the American Medical Association; various cumulative lists, nomenclature statements of USAN, etc.
Services: reference
No service charges
Operation:
manual

● ***519*** *(Organization)*
Weidner Communications Inc.
WCI
1673 West 820 North — Phone: (801) 375–9910
Provo, Utah, 84601 — Telex: 453074
USA
Staff: 17
Head: Mr. Nevil Garrett, President
Staff categories:
2 terminologist(s) (theory and special field); 3 information specialist(s); 1 translator(s); 1 terminology teacher(s); 1 documentalist(s)/librarian(s); 6 terminologist(s) (theory); 3 other professional(s)
Founded: 1977
Activities: Basic maintenance, publication and translators support capabilities; continuing development of advanced translators aids
Documentation:
bibliographic data
factographic data
Subject(s): all fields
UDC 0/9
Language(s): En; Fr; De; Es; Pt; Ar; Ja
Vocabularies: Initial development; attention to grammatical coding
Subject(s): all fields
UDC 0/9
Language(s): En; Fr; De; Es; Pt; Ar; Ja (Readily extendible to any language with Roman alphabet)
Theory: Emphasis on universal terminology format capable of accommodating entries from diverse existing term banks and glossaries; definition of intrinsic and extrinsic data in term fiches
Publications: Terminology maintenance, specialized glossaries, text-related glossaries, interactive translation aids
Services: Terminology system available alone or as adjunct to Weidner computer-assisted translation system. Both systems complemented by word processor and various input and output options, including interface to phototypesetting; translation; Current awareness facility supported in user software; other: Weidner systems also available through time-sharing on company computers
Service charges: Systems available through lease or purchase
User aids: In addition to training and documentation, users are provided with continuing consultation and diagnostic support
Operation:
computer-assisted
Data base description: See No. 520

● ***520*** *(Data Bank)*
Weidner Communications Inc.
WCI
1673 West 820 North — Phone: (801) 375–9910
Provo, Utah, 84601 — Telex: 453074
USA
Staff: 17
Head: Mr. Nevil Garrett (President)
Staff categories:
2 terminologist(s) (special field) / linguist(s); 3 information specialists; 1 translator; 1 terminology teacher; 1 documentalist / librarian; 6 terminologists (theory); 3 other professional(s)
Founded: 1977
Purpose of data bank/users: Support for technical translation; Translators, revisors, terminologists, publishers
Operation: computer-assisted
Data base description:
Name: Weidner Term Bank
Size: Minimum of 5000 core terms per language pair
Language(s): En; Fr; De; Es; Pt; Ar; Ja (Readily extendible to any language with Roman alphabet)
Classification: User-specifiable subject labels
Subject(s): all fields
UDC 0/9
Language(s): En; Fr; De; Es; Pt; Ar; Ja
Terminological data: Preparation of general and technical term banks; brokering of technical terminology. Development of maintenance and publication utilities and variety of machine aids to translation;
definitions; explanations (included with definitions as necessary); context; equivalents (available indirectly); other data elements (origin, entry date, modification date, subject field, quality rating, familiarity rating, grammatical codes, and userdefined documentation lexical relations, dialects, usage, general comments, etc.)
Products: Terminology maintenance, specialized glossaries, text-related glossaries, interactive translation aids
Only product thus far is dictionary (general language) used in computer for field users; magnetic tapes, disks (used in terminology transfer, software updates)
Services: Current awareness facility supported in user software
Service charges: Terminology system available through lease or purchase
Access: Access limited to Weidner customers; Dial-up time-sharing on company computers
on-line (full capabilities available with installed systems)
Data available: Subject to negotiation
on exchange basis; with charge
Data processing:
Hardware/Capacity: Digital Equipment Corp. PDP 11/70; 128 kb--1 Mb; 60 Mb er language pair; 800 or 1600 bpi; ANSI-standard or unlabled, fixed-block, unspanned-record; CRTs, optical scanners, line printers, character printers.; 32 ports (RS-232 protocol)
Software: Applications programs developed in-house; operating system: RSX-11M, RSX-11M+; operation mode: batch; demand; programming language(s): FORTRAN; MACRO 11; data base management system: DBMS developed in-house

○ ***521*** *(Organization)*
Associação Brasileira de normas técnicas
[Brazilian Association of Technical Standards]
ABNT

Av. 13 de Maio, Un 13–28U andar
Caixa Postal 1680 Phone: 55 21 220 44 55
Rio de Janeiro-RJ, 20. 003
Brazil Telegr.: normatécnica rio
Head: Mr. A. M. Martinez, Secretário Executivo
Vocabulary standards:
Subject(s): all fields
UDC 0/9
Language(s): Pt

◐ ***522*** *(Organization)*
Instituto de pesquisas tecnológicas do Estado de São Paulo S. A.
[Institute for Technological Research (not official)]
IPT
Cidade Universitária
P. O. Box 7141 Phone: (011) 268-2211
São Paulo, 05508 Telex: (011) 22831 INPT BR
Brazil
Head: Dr. Alberto Pereira de Castro
Founded: 1899
Vocabularies: Restricted activities at the moment
Subject(s): applied sciences; chemistry
UDC 54; UDC 6
Language(s): En; Fr; De; Pt
Publications: WAENY, J. C. C. *Recomendaçoes para nomenclatura de confiabilidade metrológica.* 1979, 23 p. (Publ.IPT no.1147.Série Pl 2.)
Services: reference; loan; reproduction
Service charges
Operation:
computer-assisted
Data processing:
Hardware: Burroughs; B-1700; 128 Kb; 260 Mb; 3 PE; 2 printers (1100 LPM, 750 LPM)
Software: operating system: MCP II; operation mode: batch; programming language(s): COBOL; Fortran

(Project(s))
Title: Terminologia em um Instituto de Pesquisa Tecnológica [Terminology in a Technical Research Institute]
Project type: institutional; in planning
Collaborators: 5
Scope and/or subject matter: To establish a controlled language for retrieving and disseminating information
Method of work: Documents are analysed and converted into a controlled vocabulary with the aid of a thesaurus. Indexing of documents.
Source material: Glossaries, dictionaries, etc. Languages covered are En, Fr, De, Pt

◐ ***523*** *(Commission/Committee)*
Grupo de trabajo sobre lenguajes de indización
[Working Group on Indexing Languages (not official); Groupe de travail sur langages d'indexation]
Parent organization: Centro Argentino de Información Cientifica y Tecnológica (CAICYT)
Moreno 431 Phone: 34-1777
Buenos Aires, 1091 Telex: 022414
Argentina
Contact: Ms. Mónica Allmand de Gallo
Capacitación y Estudios
CAICYT
Committee members: 13 (The Working Group is composed of information specialists from several institutions)
Chairperson: Ms. Mónica Allmand de Gallo (coordinator)
Established: 1978
Vocabularies:
Subject(s): indexing
UDC 025. 4
Language(s): Es
Publications: Kit on an indexing language course: Thesaurus construction. Glossary (in print by Unesco); Inventory of non-traditional documentary languages in Es and/or Pt, 1981.

○ ***524*** *(Organization)*
Instituto nacional de normalización
[National Institute of Standardization]
INN
Matias Cousino 64 - 6U piso
Casilla 995 - Correo 1 Phone: 6 81 44
Santiago Telex: 260 stgoch
Chile Telegr.: inn
Head: Mr. Pedro Vilaseca P.
Vocabulary standards:
Subject(s): all fields
UDC 0/9
Language(s): Es

○ ***525*** *(Organization)*
Dirección general de normas y tecnología. Ministerio de industria, Comercio y Turismo
[General Office of Standards and Technology. Ministry of Industry, Commerce and Tourism]
Piso 9
Casilla 4430
La Paz
Bolivia
Vocabulary standards:
Subject(s): all fields
UDC 0/9
Language(s): Es

● ***526*** *(Organization)*
Instituto de geologia minero metalúrgica
[Geological, Mining and Metallurgical Institute]
INGEMMET
Calle Pablo Bermudez, NU 211
Lima-14
Peru
Expert(s): Ing. Samame Boggio

(Project(s))
Responsible: Mining Standardization Division
Title: Glosario de Términos Mineros [Glossary of Mining Terms]
Project type: institutional; in progress
Principal investigator: Ing. Juan Guillermo Hercilla González
Collaborators: 44
Scope and/or subject matter: To develop a glossary that covers the standardization of terms uysed in the mining field
Method of work: Examination of publications from different institutions of the country, concerning the existing terminology and definitions; establishment of a commission with individuals concerned with the mining, metallurgical and geological field; visit to different mines of the country in order to get acquainted with the terms most commonly used in mining
Source material: (1) "Perú Minero" by Mario Samamé Boggio, 1979; (2) "Glossary of Mining Geology" by E. C. Amstutz, Heidelberg 1970; (3) "Minería" Journal of the Instituto de Ingenieros de Minas del Perú

○ ***527*** *(Organization)*
Instituto de investigación tecnológica industrial y de normas técnicas
[Institute of Technical Industrial Research and Technical Standards]
ITINTEC
Jr. Morelli - 2da. cuadra, Urbanización San Borjá - Surquillo Phone: 51 14 40 10 40
Lima 34 Telex: 20496 peitintec
Peru Telegr.: itintec
Head: Mr. Leonidas del Valle Bernuy
Vocabulary standards:
Subject(s): all fields
UDC 0/9
Language(s): Es

● *528* *(Organization)*
Universidad nacional de Trujillo. Departamento de idiomas y linguistica
[National University of Trujillo. Department of Second Languages and Linguistics]
UNT - d-Idl
Ciudad Universitaria
P. O. Box 315 Phone: 24–28–61
Trujillo
Peru
Staff: 20
Head: Prof. Aníbal Sánchez, Chairman of Department, until May 31, 1983
Expert(s): No official expert, but contact: Prof. Ernesto Zierer
Staff categories:
4 linguist(s); 4 translator(s); 1 terminology teacher(s)
Founded: Of the University: 1824
Documentation: bibliographic data
Subject(s): social sciences
UDC 3
Language(s): Es; En; Fr; De
Vocabularies:
Subject(s): social sciences
UDC 3
Language(s): Es; En; Fr; De
Training: Occasionally; undergraduate
Access to training: available to outside users
Curriculum: fundamentals of terminology; terminological principles
Services: reference; reproduction (limited)
No service charges
Terminological data:
definitions; equivalents

○ *529* *(Organization)*
Instituto Colombiano de normas técnicas
[Colombian Institute of Technical Standards]
ICONTEC
Carrera 37 No. 52–95
P. O. Box 14237 Phone: 255 70 55
Bogotá
Colombia Telegr.: icontec
Head: Mr. Javier-Henao L., Director Ejecutivo
Vocabulary standards:
Subject(s): all fields
UDC 0/9
Language(s): Es

○ *530* *(Organization)*
Instituto Ecuatoriano de normalización
[Standards Institute of Ecuador]
Av. Universitaria 784
Casilla 3999
Quito
Ecuador
Vocabulary standards:
Subject(s): all fields
UDC 0/9
Language(s): Es

● *531* *(Data Bank)*
Banco de datos terminológicos de la Universidad Simón Bolivar
[Terminological data bank of the University Simón Bolivar]
BTUSB
Parent organization: Universidad Simón Bolívar
P. O. Box 1512
Caracas, 1010A
Venezuela
Head: Prof. D. Raventos de Castro
Founded: in planning
Purpose of data bank/users: The bank will serve as a source of reference for teachers, researchers and national organisms.
Subject(s): technical fields (mainly)
UDC 6
Language(s): Es; En; Fr
Terminological data: The bank is to cover all fields dealt with at the University of Bolivar

○ *532* *(Organization)*
Comisión Venezolana de normas industriales
[Venezuelan Commission of Industrial Standards]
COVENIN
Avda. Andrés Bello, Edif. Torre Fonde Comun, Pisos 11, 12 y 13 Phone: 575 41 11
Caracas 1050 Telex: 22753
Venezuela Telegr.: covenindus
Head: Dr. Jesuardo E. Areyan Salazar, Executive Secretary
Vocabulary standards:
Subject(s): all fields
UDC 0/9
Language(s): Es

○ *533* *(Organization)*
Instituto Uruguayo de normas técnicas
[Uruguay Institute of Technical Standards]
Av. Libertador Brig. Gral., Lavallaja 1464 P. 9
Montevideo
Uruguay
Vocabulary standards:
Subject(s): all fields
UDC 0/9
Language(s): Es

○ *534* *(Organization)*
Badan Kerjasama Standardisasi LIPI-YDNI
[LIPI-YDNI Joint Standardization Committee]
YDNI
Jln. Teuku Chik Ditiro 43
P. O. Box 250 Phone: 35 16 58
Jakarta
Indonesia Telegr.: lipi jakarta
Vocabulary standards:
Subject(s): all fields
UDC 0/9
Language(s): In

● *535* *(Organization)*
Pusat Dokumentasi Ilmiah Nasional
[National Scientific Documentation Center]
PDIN
Parent organization: Lembaga Ilmu Pengetahuan Indonesia (Indonesian Institute of Sciences)
Jalan Gatot Subroto, Kompleks LIPI
P. O. Box 3065/Jkt Phone: 583465; 511063
Jakarta Telex: 45875 IA
Indonesia Telegr.: PDIN
Staff: 4
Head: Miss Luwarsih Pringgoadisurjo, M. A.
Expert(s): Zultanawar, M. A.
Staff categories:
3 documentalist(s)/librarian(s); 1 linguist(s)
Founded: 1965
Documentation:
bibliographic data
factographic data
Subject(s): all fields
UDC 0/9
Language(s): In; En
Vocabularies:
Subject(s): library science; documentation; information
UDC 02; UDC 007
Language(s): In; En
Vocabulary standards:
Subject(s): library science; documentation; information
UDC 02; UDC 007
Language(s): In
Publications: BACA (READ), a bi-monthly newsletter, published since 1974; Index of Indonesian learned

periodicals (semi-annually); Index to research and survey reports (annually); Directory of special libraries in Indonesia, 1978 (up-dated regularly); Special bibliographies (Published on a continuing basis)
Services: reference; loan; reproduction; translation; referral
User aids: Newsletter
Operation:
manual
Access:
via telephone / via telex

(Project(s))
Responsible: Department of Mines and Energy
Title: Thesaurus bidang pertambangan non-minyak [Thesaurus of mining terms excluding petroleum]
Project type: institutional; in progress
Principal investigator: Mr. R. A. Sunardi
Collaborators: 6
Scope and/or subject matter: Mineral resources; mineral engineering; mineral commodities; related fields (management, information, finance, etc.)
Method of work: Selecting terms from several thesauri, lists of subject headings, indexes, etc. in the field of engineering, geology, mining, management, information, etc. Reviewing and adding new terms by specialists in the subjects

○ ***536*** *(Organization)*
Product Standards Agency. Ministry of Trade and Industry
PSA
361 Buendia Avenue Extension Phone: 86 66 86
Makati, Metro Manila
Philippines Telegr.: philstand manila
Head: Mr. I. G. Salcedo, Director
Vocabulary standards:
Subject(s): all fields
UDC 0/9
Language(s): En

● ***537*** *(Organization)*
Surian Ng Wikang Pambansa
[Institute of National Language]
SWP; INL
Parent organization: Ministry of Education and Culture
Arroceros Street
P. O. Box 2287 Phone: 49 29 24; 49 35 68
Manila, 2801
Philippines Telegr.: National Language, Arroceros
Staff: 136
Head: Atty. Ponciano B. P. Pineda, Ph. D.
Staff categories:
2 terminologist(s) (special field); 40 terminologist(s) (theory); 1 documentalist(s)/librarian(s); 15 linguist (s); 5 information specialist(s); 6 translator(s)
Founded: 1936
Documentation:
bibliographic data
factographic data
Subject(s): all fields
UDC 0/9
Language(s): En; Pi
Vocabularies:
Subject(s): all fields
UDC 0/9
Language(s): En; Pi
Services: translation
User aids: Newsletter
Data available:
free of charge

(Project(s))
Title: Mga Katawagan sa Edukasyong Bilinggwal [Terms in Bilingual Education]
Project type: institutional; in progress
Principal investigator: Atty. Ponciano B. P. Pineda
Collaborators: 10
Scope and/or subject matter: Terms in Bilingual Education. The book contains 18. 000 terms commonly used in bilingual education. It is intended to meet the needs of teachers handling social studies/social science, work education, character education, health education and physical education in Pilipino.
Method of work: Gathering and indexing method
Source material: Term papers, theses, mimeograph, actual research articles

Title: Mga Katawagang Panghanapbuhay [Occupational Terms]
Project type: institutional; in progress
Principal investigator: Atty. Ponciano B. P. Pineda
Collaborators: 10
Scope and/or subject matter: Occupational Terms. The book lists 18 083 terms English with 58 756 Pilipino equivalents in agriculture, vocational/industrial education, office management, fishing and home industries.
Method of work: Gathering and indexing method
Source material: Term papers, theses, mimeographs, books, actual research articles

Title: Mga Katawagang sa Pananaliksik [Research Terminologies]
Project type: institutional; in progress
Principal investigator: Atty. Ponciano B. P. Pineda
Collaborators: 10
Scope and/or subject matter: This is a compilation of terms in research with corresponding equivalents in Pilipino. There are 2 879 English terms with 4 634 equivalents. The terms which are arranged alphabetically were taken from the term papers, theses, mimeographs, books, research, statistics and from actual research articles
Method of work: Gathering and indexing method
Source material: Term papers, theses, mimeographs, actual research articles

Title: Mga Katawagang Teknikal [Technical Terms]
Project type: institutional; in progress
Principal investigator: Atty. Ponciano B. P. Pineda
Collaborators: 10
Scope and/or subject matter: Technical Terms. Bilingual technical terms in electronics, electricity, automotive engineering
Method of work: Gathering and indexing method
Source material: Term papers, theses, mimeographs, books, actual research articles

○ ***538*** *(Organization)*
Standards Association of New Zealand
SANZ
Private Bag Phone: 64 4 842 108
Wellington Telex: 3850 sanz
New Zealand Telegr.: standards
Head: Mr. Denys R. M. Pinfold, Director
Vocabulary standards:
Subject(s): all fields
UDC 0/9
Language(s): En

○ ***539*** *(Organization)*
Standards Association of Australia
SAA
80–86 Arthur Street -
Standards House Phone: 61 2 929 60 22
North Sydney, N. S. W. 2060 Telex: 26514 astan
Australia Telegr.: austandard north sydney
Head: Mr. J. R. Paton, Director General
Vocabulary standards:
Subject(s): all fields
UDC 0/9
Language(s): En

9*

ABNT	Associação brasileira de normas técnicas
ACC	Administrative Committee on Coordination (of UN)
ACLA	Association canadienne de linguistique appliquée
AFNOR	Association française de normalisation
AFTERM	Association française de terminologie
AGI	American Geological Institute
AIED	Association internationale d'entreprises de dragage
AILA	Association internationale de linguistique appliquée
AIPCR	Association internationale permanente des congrès de la route
AITT	Association internationale de traitement thermique des matériaux
ALECSO	Arab League Educational, Cultural and Scientific Organization
AMA	American Medical Association
ANSI	American National Standards Institute
ARCSS	The Arab Region Center for Social Science Research and Documentation
ARSO	African Regional Standards Organization
ASMO	Arab Standardization and Metrology Organization
ASMW	Amt für Standardisierung, Meßwesen und Warenprüfung
ATA	American Translators Association
ATT	Agnew Tech-Tran Inc.
AVT/TD	Hoofafdeling vertalingen / Bureau terminologie en documentatie
BAST	Bundesanstalt für Straßenwesen
BCS	Bureau of Ceylon Standards
BGR	Bundesanstalt für Geowissenschaften und Rohstoffe
BHRA	British Hydromechanics Research Association
BIPM	Bureau international des poids et mesures
BIRD	Banque internationale pour la reconstruction et le développement
BIRF	Banco internacional de reconstrucción y fomento
BISFA	Bureau international pour la standardisation de la rayonne et des fibres synthétiques
BLLD	British Library Lending Division
BNQ	Bureau de normalisation du Québec
BNSI	Barbados National Standards Institution
BRGM	Bureau de recherche geólogique et minière
BSI	British Standards Institution
BSprA	Bundessprachenamt
BTUSB	Banco de datos terminologicos de la Universidad Simón Bolívar
CAICYT	Centro argentino de información científica y tecnológica
CAS	China Association for Standardization
CATN	Comisión asesora de terminología nuclear
CBE	Committee of Biology Editors
CCE	Commission des communautés européennes
CCE BTB	Commission des communautés européennes. Bureau de terminologie de Bruxelles
CCE-BTL	Commission des communautés européennes. Bureau de terminologie (Luxembourg)
CCEP	Conseil consultatif des études postales
CCL-UMIST	Centre for Computational Linguistics. University of Manchester Institute of Science and Technology
CCPS	Consultative Council for Postal Studies
CDU	Classification décimale universelle
CEBA	Copenhagen School of Economics and Business Administration
CEC	Commission of the European Communities
CEC BTB	Commission of the European Communities. The Brussels Terminology Bureau
CEC-BTL	Comission of the European Communities. Bureau of Terminology (Luxemburg)
CECIP	Comité européen des constructeurs d'instruments de pesage
CEE	Communauté économique européenne / Commissions économiques européennes
CEI	Commission électrotechnique internationale
CEI/CE	Comission électrotechnique internationale. Comité d'étude
CEMT	Conférence européenne des ministres des transports
CEN	Comité européen de coordination des normes
CENELEC	Comité européen de coordination des normes électriques / Committee for the Coordination of European Standards in the Electrical Field
CETTF	Comité d'études des termes techniques français
CGE	Compagnie générale électrique du Canada
CIE	Commission internationale de l'éclairage
CIIA	Commission internationale des industries agricoles et alimentaires
CILF	Conseil international de la langue française
CINZ	Commission internationale de nomenclature zoologique
CIOMS	Council for International Organizations of Medical Sciences
CIPM	Comité international des poids et mesures
CIRP	Collège international pour l'étude scientifique des techniques de production mécanique
CISS	Conseil international des sciences sociales
CITT	Comité interentreprises de traitement de textes
CMEA	Council for Mutual Economic Aid
CNDST-BR	Centre national de documentation scientifique et technique, Bibliothèque Royale Albert I
CNGI	Comité des normes gouvernementales en informatique
CNOC	Commission on Nomenclature of Organic Chemistry
CNRS	Centre national de recherche scientifique
COCTA	Committee on Conceptual and Terminological Analysis
CODATA	Committee on Data for Science and Technology
COGEODATA	Committee on Storage, Automatic Processing and Retrieval of Geological Data
COISM	Conseil des organisations internationales des sciences médicales

COLT	Bureau of Conferences, Languages and Documents Translation
COMTERM	Commission de terminologie / Commission on Terminology
COPANT	Comisión panamericana de normas técnicas
COSQC	Central Organization for Standardization and Quality Control
COT	Committee on Terminology
COVENIN	Comisión venezolana de normas industriales
CPDT	Centre de préparation documentaire à la traduction
CRB	Centre suisse d'études pour la rationalisation du bâtiment
CRC	Computing Research Centre
CSIR	Council for Scientific and Industrial Research
CSK	Committee for Standardization of the Democratic People's Republic of Korea
CT	Comité technique
CT ASAS	Commission terminologique. Académie des sciences et des beaux arts de Slovénie
CTB	Centre de terminologie de Bruxelles / Centrum voor terminologie van Brussel
CTFA	Cosmetic, Toiletry and Fragrance Association
CTOLF	Commission de terminologie de l'Office de la langue française
CUF	Conseil de l'Union française
CUTA	Comité d'uniformisation de la terminologie aéronautique
CUTA (ATC)	Comité d'uniformisation de la terminologie aéronautique (contrôle de la circulation aérienne)
DDC	Dewey Decimal Classification
DES	Department of Education and Science
DGD	Deutsche Gesellschaft für Dokumentation, e. V.
DGN	Dirección general de normas
DGQ	Direcção-geral da qualidade
DGTD	Direction générale de la terminologie et de la documentation
DIGENOR	Dirección general de normas y sistemas de calidad
DIN	DIN Deutsches Institut für Normung e. V.
DIN-NAT	Deutsches Institut für Normung e. V., Normenausschuß Terminologie
DIRR	Direction générale de recherche routière
DKS	D"rzaven komitet za standardizacija pri ministerskija s"vet
DS	Dansk Standardiseringsraad
DUSH	Društva za unapredenja standardizacije Hrvatske
EC	European Community
ECA	Economic Commission for Africa (UN)
ECE	Economic Commission for Europe (UN)
ECLA	Economic Commission for Latin America (UN)
ECMT	European Conference of Ministers of Transport
ECOR	Engineering Committee on Oceanic Resources
EEC	European Economic Community
ELOT	Ellinikos organismos typopiiseos
EM	Excerpta Medica
EOS	Egyptian Organization for Standardization
EP	European Parliament / Europa-Parlamentet / Europäisches Parlament / Europees Parlement
ERIC	Educational Resources Information Center
ESA	European Space Agency
ESCAP	Economic and Social Commission for Asia and the Pacific (UN)
ESP	English for Special Purposes
ETI	Ecole de traducteurs et d'interprètes
EUMABOIS	Comité européen des constructeurs de machines à bois
FAO-TERM	Food and Agriculture Organization of the United Nations. Terminology and Reference Section
FEANI	Fédération européenne d'associations nationales d'ingénieurs
FHB	Fachhochschule für Bibliothekswesen Stuttgart
FIACC	Five International Associations Coordinating Committee
FID	Fédération internationale de documentation
FIT	Fédération internationale des traducteurs
FIZ 5	Fachinformationszentrum "Hüttenkunde, Werkstoffe, Metallbe- und -verarbeitung"
FLACSO	Facultad latinoamericano de ciencias sociales
FMI	Fonds monétaire international / Fondo monetario internacional
GAL	Gesellschaft für Angewandte Linguistik
GATT	General Agreement on Tariffs and Trade
GESC	Government EDP Standards Committee, Task Force on Terminology
GETC	Groupe d'étude de la terminologie comptable
GID	Gesellschaft für Information und Dokumentation
GIRSTERM	Groupe interdisciplinaire de recherche scientifique et appliquée en terminologie
GOST	Gosudarstvennyj komitet standartov SSSR
GSB	Ghana Standards Board
GT	Groupe de travail
GTZ	Deutsche Gesellschaft für technische Zusammenarbeit
HELPIS	Higher Education Learning Programmes Information Service
HIVT	Hoger instituut voor vertalers en tolken van het Rijksuniversitair Centrum Antwerpen
HKSTC	Hong Kong Standards and Testing Centre
IADC	International Association of Dredging Companies
IAEA	International Atomic Energy Agency
IAPH	International Association of Ports and Harbours
IBE	International Bureau of Education
IBLV	Institut Bourguiba des langues vivantes
IBM	International Business Machines
IBN	Institut belge de normalisation

IBRD	International Bank for Reconstruction and Development
ICA	International Cartographical Association
ICA	International Council on Archives
ICAO	International Civil Aviation Organization
ICC	International Association for Cereal Chemistry / Association internationale de chimie céréalière
ICEC	International Cryogenic Engineering Conference
ICES	International Council for the Exploration of the Sea
ICOLD	International Commission on Large Dams
ICONTEC	Instituto colombiano de normas técnicas
ICP	International Commission for Palynology
ICS	International Commission on Stratigraphy
ICSTI	International Centre for Scientific Information
ICSU	International Council of Scientific Unions
ICU	International Cartography Union
ICUTN/WBIT	International Committee for Unification of Terminological Neologisms. World Bank of International Terms
ICYT	Instituto de información y documentación en ciencia y tecnología
ICZN	International Commission on Zoological Nomenclature
IDA	International Development Association
IDCNS	Interdivisional Committee on Nomenclature and Symbols
IDF	International Dairy Federation
IEC	International Electrotechnical Commission
IFC	International Finance Corporation
IFHT	International Federation for the Heat Treatment of Materials
IFIP	International Federation for Information Processing
IFLA	International Federation of Library Associations
IGCP	International Geological Correlation Programme
IGT	Institute of Gas Technology
IGU	International Gas Union / Internationale Gasunion
IGU	International Geographical Union
IIF	Institut international du froid
IINTE	Instytut informacji naukowej, technicznej i ekonomicznej
IIR	International Institute of Refrigeration
IIRS	Institute for Industrial Research and Standards
IIS	Institut international de la soudure
IIW	International Institute of Welding
IKR	Institute of Kiswahili Research
ILO	International Labour Organization
IMCO	Intergovernmental Maritime Consultative Organization (UN)
IMECO	International Measurement Confederation
IMF	International Monetary Fund
IMIC	International Medical Information Centre
INAPI	Institut algérien de normalisation et de propriété industrielle
INFOTERM	International Information Centre for Terminology / Centre international d'information pour la terminologie / Internationales Informationszentrum für Terminologie / Meždunarodnyj informacionnyj centr po terminologii
INGEMMET	Institutio de geología minero metalúrgica
INIV	Institut für nichtnumerische Informationsverarbeitung
INN	Instituto nacional de normalización
INTERGEO	Laboratoire d'information et de documentation en géographie "INTERGEO"
IOB	Inter-Organisation Board for Information Systems
IOHE	International Organization for Human Ecology
IOLM	International Organization of Legal Metrology
IPA	International Pharmaceutical Abstracts
IPT	Instituto de pesquisas tecnológicas do Estado de São Paulo S. A.
IRANOR	Instituto nacional de racionalización y normalización
IRN / CT	Instituto nacional de racionalización y normalización, Comisión de trabajo
IRRC	International Road Research Centre
IRRD	International Road Research Documentation
IRS	Institutul român de standardizare
IRU	International Railway Union
ISI	Indian Standards Institute
ISIRI	Institute of Standards and Industrial Research of Iran
ISO	International Organization for Standardization / Organisation internationale de normalisation / Meždunarodnaja organizacija po standartizacii
ISTEI	Institute for Scientific, Technical and Economic Information
ITC	Stichting International Translation Centre
ITINTEC	Instituto de investigación tecnológica industrial y de normas técnicas
ITU	International Telecommunication Union
IUB	International Union of Biochemistry
IUCr	International Union of Crystallography
IUFOST	International Union of Food Sciences and Technology
IUGS	International Union of Geological Sciences
IUH	International Union of Suction Hopper Dredger Owners
IUPAC	International Union of Pure and Applied Chemistry
IUPAP	International Union of Pure and Applied Physics
IUPS	International Union of Physiological Sciences
IVW	Internationaler Verband für die Wärmebehandlung der Werkstoffe

JAPIC	The Japan Pharmaceutical Information Centre
JBS	Jamaica Bureau of Standards
JCBN	Joint Commission on Biochemical Nomenclature
JCG	Joint Coordination Group
JEN	Junta de energía nuclear
JICST	The Japan Information Center of Science and Technology
JISC	Japanese Industrial Standards Committee
JISI	Junta interorganizacional para sistemas de información
JMLA	Japan Medical Library Association
JSME	The Japan Society of Mechanical Engineers
JULS	Jazykovedný ústav Ľudovíta Štúra
JZS	Jugoslovenski zavod za standardizcaciju
KBS	Bureau of Standards (Rep. of Korea)
KEBS	Kenya Bureau of Standards
KEME	Kéndron ekpedeftikón meletón ké epimorfóseos
KNJ	Katedra neslovanských jazyku
KT PAN	Komitet terminologii. Polska Akademia nauk
KTS	Komitee für Terminologie und Sprachfragen der DGD
LAD	Division of the Unesco Library, Archives and Documentation Services
LCSH	Library of Congress Subject Headings
LEAT	Legal, Economic and Administrative Terminology. Institute for Comparative Terminology
LIBNOR	Lebanese Standards Institution
MANU	Makedonska akademija na naukite i umetnostite
MEDIS	The Medical Information System Development Center
MOTOM	Meždunarodnoe obščestvo po termičeskoj obrabotke materialov
MSZH	Magyar szabványügyi hivatal
MTKL	Medzinárodná komisia pre lingvistickú terminológiu
NC	Comité estatal de normalización
NC	Nomenclature Committee
NCI	National Cancer Institute
NLM	National Library of Medicine
NNI	Nederlands normalisatie-instituut
NORMATERM	Normalisation - Automation de la terminologie
NOTE	Norsk termbank
NSF	Norges standardiseringsforbund
NSO	Nigerian Standards Organisation
OACI	Organisation de l'aviation civile internationale
OAS	Organization of American States
OAU	Organization of African Unity
OBN	Office of Biochemical Nomenclature
OECD	Organization for Economic Cooperation and Development
OIE	Office international des épizooties
OIML	Organisation internationale de métrologie légale
OIV	Office international de la vigne et du vin
OLF	Office de la langue française
OLF-BTQ	Office de la langue française - Banque de terminologie du Québec
OMM	Organisation météorologique mondiale
OMS	Organisation mondiale de la santé
ON	Österreichisches Normungsinstitut
ONU	Organisation des Nations unies / Organización de las Naciones unidas
ONUG	Office des Nations unies à Genève
OON	Organizacija Ob"edinennyh nacii
OTETO	Omádha tehnikís ké tehnologhikís orologhías
OTSZK	Országos tervhivatal számítástechniaki központja
OUA	Organisation de l'unité africaine
PDIN	Pusat dokumentasi ilmiah nasional
PE	Parlamento europeo / Parlement européen
PIANC	Permanent International Association of Navigation Congresses
PIARC	Permanent International Association of Road Congresses
PIRA	Printing Industry Research Association (U. K.)
PKNiM	Polski komitet normalizacij i miar
PKTE	Polski komitet terminologii elektrycznej
PNUE	Programme des Nations unies pour l'environnement
PNUMA	Programa de las Naciones unídas para el medio ambiente
PSA	Product Standards Agency
PSI	Pakistan Standards Institution
RC	Referalni centar Sveučilišta u Zagrebu / Referral Centre of the University of Zagreb
RTT	Rådet for teknisk terminologi
SAA	Standards Association of Australia
SABS	South African Bureau of Standards
SAITINT	The South African Institute of Translators and Interpreters
SAIVERT	Die Suid-Africaanse vertalerinstituut
SAIW	Suid-Africaanse inligtinsentrum vir water
SANZ	Standards Association of New Zealand
SASA	Slovene Academy of Sciences and Arts
SAWIC	South African Water Information Centre
SAZU	Slovenska akademija znanosti in umetnosti
SC	Sub Committee

SCAR	Scientific Committee on Antarctic Research
SCC	Standards Council of Canada
SCERGIE	Société civile d'étude et de recherche en gestion des informations de l'entreprise
SEAMIC	Southeast Asian Medical Information Center
SERC	Science and Engineering Research Council
SFS	Suomen Standardisoimisliitto r. y.
SGAT	Study Group on Accounting Terminology
SII	Standards Institution of Israel
SIPM	Shell internationale petroleum maatschappij B. V.
SIRIM	Standards and Industrial Research Institute of Malaysia
SIS	Standardiseringskommissionen i Sverige
SISIR	Singapore Institute of Standards and Industrial Research
SMTWL	Stichting voor moeilijk toegankelijke wetenschappelijke literatuur
SNIMA	Service de normalisation industrielle marocaine
SNV	Association suisse de normalisation
SP	Secrétariat pilote
SSDC	Social Science Documentation Centre (Unesco)
STC	Society for Technical Communication
STQ	Société des traducteurs du Québec
SUN-AMCO	Commission on Symbols, Units, Nomenclature, Atomic Masses and Fundamental Constants
SWP	Surian Ng Wikang Pambansa
TC	Technical Committee
TDB	Terminology and Documentation Branch
TEAM	Terminologie-Erfassungs- und Auswertungs-Methode / Terminology Evaluation and Acquisition Method
TERMIUM	La banque de terminologie du Gouvernement canadien / Canadian Government Terminology Bank
TISI	Thai Industrial Standards Institute
TK	Terminologiská komisia
TLF	Trésor de la langue française
TNC	Tekniska nomenklaturcentralen
TRRL	Transport and Road Research Laboratory
TSA	Terminologieverenigung van Suid-Afrika / Terminology Association of South Africa
TSE	Türk standardlari enstitüsü
TSK	Tekniikan sanastokeskus
TUKI	Taasisi ya uchunguzi wa Kiswahili
UAM	Uniwersytet im. Adama Mickiewicza
UATI	Union des associations techniques internationales
UDC	Universal Decimal Classification
UIC	Union internationale des chemins de fer
UIIG	Union internationale de l'industrie du gaz
UIT	Union internationale des télécommunications
UJC	Ústav pro jazyk ceský
UN	United Nations Organization
UNCTAD	United Nations Conference on Trade and Development
UNDP	United Nations Development Programme
UNEP	United Nations Environment Programme
UNESCO	United Nations Educational, Scientific and Cultural Organization
UNHCR	United Nations High Commission for Refugees
UNI	Ente nazionale italiano di unificazione
UNICEF	United Nations Children's (Emergency) Fund
UNIDO	United Nations Industrial Development Organization
UNISIST	Intergovernmental Programme for Cooperation in the Field of Scientific and Technological Information
UNITAR	United Nations Institute for Training and Research
UNM	Úrad pro normalizaci a méréni
UNOG	United Nations Office at Geneva
UNRWA	United Nations Relief and Works Agency
UNT dIdl	Universidad nacional de Trujillo. Departamento de idiomas y lingüística
UNU	United Nations University
UPU	Union postale universelle / Universal Postal Union
USAN	United States Adopted Names Council
UVTEI	Ústředí védeských, technických a ekonomických informací
VCP	Vsesojuznyj centr perevodov naučno-tehničeskoj literatury i dokumentacii Gosudarstvennogo komiteta SSSR po nauke i tehnike i Akademii nauk SSSR
VDI	Verein Deutscher Ingenieure
VNIIKI	Vsesojuznyj naučno-issledovatel'skij institut tehničeskoj informacii, klassifikacii i kodirovanija
VVS	Výskumné výpočtové stredisko
WCI	Weidner Communications
WFC	World Food Council
WFEO	World Federation of Engineering Organizations
WFP	World Food Programme
WG	Working Group
WHO	World Health Organization
WIPO	World Intellectual Property Organization
WMO	World Meteorological Organization
WNT	Wydawnictwa naukowo-techniczne
WODA	World Dredging Association
YDNI	Badan Kerjasama Standardisasi LIPI-YDNI

Worksheet 1
Organisations

001 Reference number
002 Date of record
003 Ref. no(s). of project(s) undertaken
005 Ref. no(s). of commissions(s)
006 Ref. no. of data bank
010 A
100 Name of organization (original language)
101 Translation into English (if in use)
102 Translation into French (if in use)
103 Translation into other language(s) (if in use)
110 Name of parent organizations (if applicable)
120 Acronym (original)
121 Acronym E (if in use)
122 Acronym F (if in use)
123 Acronym (other languages)
130 Address (country/state)
131 Zip (postal) code (if applicable)
132 City
133 Street/Block
134 P. O. Box
135 Telephone
136 Telex
137 Telegramme address
140 Visiting address (if applicable)
150 Year of foundation/establishment
200 Information about personnel
201 Name and title of director
202 Name of expert (official) responsible for terminology work
210 Total number of staff
211 Number of professional staff
212 Number of general staff
220 Staff categories
- G = terminologist specializing in theory/principles of terminology
- S = terminologist working in a specific field
- D = terminological documentalist/librarian, reference officer
- L = linguist (applied linguistics)
- I = information specialist
- T = translator
- E = specializing in terminological teaching/training
- O = other

300 Collaboration with other organizations
301 worldwide
302 regional (specify)
303 sub-regional (specify)
304 national (specify)
305 List organizations
310 Collaboration within networks
311 List networks
400 Terminological activities
410 Documentation of terminology
- A bibliographical data (literature, documentation)
- B terminological data (fact documentation)
- C both

411 Subject scope (please be as specific as possible)
412 UDC of subject scope
413 Language(s)
414 Terminological holdings
- A books (general)
- B dictionaries (general)
- C specialized vocabularies/glossaries/word lists
- D works of reference
- E bibliographies
- F standards
- G journals
- H theses, dissertations
- I articles
- J others

415 Holdings in: A book form; B microfiche
416 Holdings in computerized form
C Cards; T Tape; D Disc
417 Indexing or classification tools used (please specify)
418 General access to holdings
419 Restricted acces (please specify)
420 Elaboration of terminologies (lexicographical work)
421 Field(s)
422 UDC of field(s)
423 Language(s)
430 Work on theory of terminology
433 Language(s)
440 Work on standardization of terminology
441 Field(s)
442 UDC of field(s)
443 Language(s)
450 Teaching of terminology
451 Level: A undergraduate, B graduate, C postgraduate
452 Available to outside users
453 Specifics about course
Course includes:
- A Fundamentals of terminology
- B Terminological principles
- C Methods of terminological lexicography

457 Elaboration of teaching material
458 Available teaching material (please specify)
460 Publications issued by your organization (please provide proper bibliographical references)
470 Services provided by your organization
471 Reference services
472 Loan services
473 Reproduction services
474 Translation services
475 Referral services
476 Other (please specify)
477 Charge(s) for services
478 No charge for services
480 User Aids
481 Newsletter
482 Others
500 Type of operation
511 Manual — Please go to 0540
512 Computer assisted
513 Fully computerized
520 Hardware
521 Make (brand name, eg. IBM)
522 Model (e.g. 370)
523 Memory (e.g. 500 Kb)
524 Storage capacity (e.g. 5 Mb)
525 Magnetic tape
526 Other onsite peripherals (e.g. printer, COM-)
527 Terminal connections (please specify)
529 Additional information (please specify)
530 Software
531 Operating system
532 Operation mode(s) (batch; real time; demand)
533 Programming language(s) (e.g. PL1)
534 Data base management system
535 Retrieval documentation system
539 Additional information (please specify)
540 Data base description
541 Name of data base
542 Size of data base
543 Update rate
544 Language(s) of data base
545 Information language (classification system, thesaurus used, eg. UDC, INIS)
546 Other indexing method(s) (please specify)
549 Additional information (please specify)
550 Concept description includes:
551 Definition of concepts
552 Explanation(s) of concepts
553 Indication of systematic representation of concepts (representation of concept relations)
554 Context
555 Standardized terms only
556 Equivalent terms in other languages

557 Other data elements
559 Additional information
560 Services and products
561 Products issued with the aid of the bank, such as glossaries, word lists, etc. (Please provide proper bibliographical refs.)
562 Microforms (please specify)
563 Magnetic tapes, discs (please specify)
564 Current awareness service (SDI)
565 Other services offered (please specify)
566 Charge for services
567 No charge for service

570 Access to data base for outside users
571 Via telephone
572 Via telex
573 Other (please specify)
574 On-line
575 Off-line
580 Availability of data to outside users
581 On exchange basis
582 With charge
583 Free of charge
584 Not available
585 Other

Worksheet 2
Commissions/Committees

001 Reference number
002 Date of record
003 Ref. no(s). of projects undertaken
004 Ref. no. of organization
010 B
700 Name of commission (in original language)
701 Translation into E
702 Translation into F
703 Translation into other language(s)
710 Name of parent organization
720 Acronym of commission
721 Acronym E (if in use)
722 Acronym F (if in use)
723 Acronym (other languages)
750 Year of establishment
800 Information about personnel
801 Name of chairperson
802 Name of vice-chairperson
810 Approximate number of committee (commission) members
830 Address of parent organization (country/state)
831 Zip (postal) code (if applicable)
832 City
833 Street/Block
834 P. O. Box
835 Telephone
836 Telex
837 Telegramme address

840 Visiting address (if applicable)
841 Name and address of person to whom inquiries concerning commission work should be sent (if other than given in 830 — 840)
900 Collaboration with other organizations
901 worldwide
902 regional (please specify)
903 sub-regional (please specify)
904 national (please specify)
905 List organizations
910 Collaboration within networks (please specify)
911 List networks
920 Elaboration of terminologies (lexicographical work)
921 Field(s)
922 Language(s)
923 UDC of field(s)
930 Work on theory of terminology
933 Language(s)
940 Work on standardization of terminology
941 Field(s)
942 UDC of field(s)
943 Language(s)
960 Publications issued by your committee/commission on activities described (please provide bibliographical refs.)
961 Charges for such publications
962 Publications without charge

Worksheet 3
Terminological data banks

001 Reference number
002 Date of record
004 Ref. no. of organization
010 C
100 Name of organization (original language)
101 Translation into English (if in use).
102 Translation into French (if in use)
103 Translation into other language(s) (if in use)
110 Name of parent organizations (if applicable)
120 Acronym (original)
121 Acronym E (if in use)
122 Acronym F (if in use)
123 Acronym (other languages)
130 Address (country/state)
131 Zip (postal) code (if applicable)
132 City
133 Street/Block
134 P. O. Box
135 Telephone
136 Telex
137 Telegramme address
140 Visiting address (if applicable)
150 Year of foundation/establishment
200 Information about personnel
201 Name and title of director

202 Name of expert (official) responsible for terminology work
210 Total number of staff
211 Number of professional staff
212 Number of general staff
220 Staff categories
G = terminologist specializing in theory/principles of terminology
S = terminologist working in a specific field
D = terminological documentalist/librarian, reference officer
L = linguist (applied linguistics)
I = information specialist
T = translator
E = specializing in terminological teaching/training
O = other
300 Collaboration with other organizations
301 worldwide
302 regional (specify)
303 sub-regional (specify)
304 national (specify)
305 List organizations
310 Collaboration within networks
311 List networks

800 Activities
811 Field(s) covered
812 UDC
813 Language(s) covered
814 Purpose of t.d.bank
815 Users (e.g. subject specialists, translators, etc.)
816 Type of operation:
A Computer assisted
B Fully computerized
820 Hardware
821 Make (Brand name, e.g. IBM)
822 Model (e.g. 370)
823 Memory (e.g. 500 Kb)
824 Storage capacity (e.g. 5 Mb)
825 Magnetic tape
826 Other onsite peripherals (e.g. printer, COM-)
827 Terminal connections (please specify)
829 Additional information (please specify)
830 Software
831 Operating system
832 Operation mode(s) (batch; real time; demand)
833 Programming language(s) (e.g. PL1)
834 Data base management system
835 Retrieval documentation system
839 Additional information (please specify)
840 Data base description
841 Name of data base
842 Size of data base
843 Update rate
844 Language(s) of data base
845 Information language (classification system, thesaurus used, eg. UDC, INIS)
846 Other indexing method(s) (please specify)
849 Additional information (please specify)
850 Concept description
851 Definition of concepts
852 Explanation(s) of concepts
853 Indication of systematic representation of concepts (representation of concept relations)
854 Context
855 Standardized terms only
856 equivalent terms in other languages
857 other data elements
859 Additional information
860 Services and products
861 Products issued with the aid of the bank, such as glossaries, word lists, etc. (Please provide proper bibliographical refs.)
862 Microforms (please specify)
863 Magnetic tapes, discs (please specify)
864 Current awareness service (SDI)
865 Other services offered (please specify)
866 Charge for services
867 No charge for services
870 Access to data bank for outside users
871 Via telephone
872 Via telex
873 Other (please specify)
874 On-line
875 Off-line
880 Availability of data to outside users
881 On exchange basis
882 With charge
883 Free of charge
884 Not available
885 Other

Worksheet 4
Projects

001 Reference number
004 Ref. no. of organization/commission undertaking project
010 D
601 Name of organization undertaking project
602 Name of commission undertaking project
603 Name of individual undertaking project
610 Title of project (original language)
611 Translation of title into E
612 Type of project
A Institutional project
B Thesis, dissertation, habilitation
C Commissioned project (please provide name of organization commissioning project)
613 Status of project
A in progress
B in planning
614 (ad A) projected date of completion
615 (ad B) projected dates/or duration
616 Number of persons collaborating on project
617 Name of director(s) or persons in charge
618 Scope of project, i. e. subject field of project (please specify)
622 Available or forthcoming publications (please provide bibliographical references)
623 Unpublished papers available on project (mimeographs, working papers, interim reports, etc.) (please provide bibliographical references)
624 Data available
C... on cards
T... on tape
D... on disc
626 Availability of data to outside users
A... on exchange basis
B... no exchange, with charge
C... free of charge
D... not available

Kolibri 360 GmbH
Ferdinand-Jühlke-Straße 7
99095 Erfurt, Deutschland
produktsicherheit@kolibri360.de